RUSSIAN TRANSLATIONS SERIES

1. K.Ya. Kondrat'ev et al. (editors): *USSR/USA Bering Sea Experiment*
2. D.V. Nalivkin: *Hurricanes, Storms and Tornadoes*
3. V.M. Novikov (editor): *Handbook of Fishery Technology*, Vol. 1
4. F.G. Martyshev: *Pond Fisheries*
5. R.N. Burukovskii: *Key to Shrimps and Lobsters*
6. V.M. Novikov (editor): *Handbook of Fishery Technology*, Vol. 4
7. V.P. Bykov (editor): *Marine Fishes*
8. N.N. Tsvelev: *Grasses of the Soviet Union*
9. L.V. Metlitskii et al.: *Controlled Atmosphere Storage of Fruits*
10. M.A. Glazovskaya: *Soils of the World* (2 volumes)
11. V.G. Kort & V.S. Samoilenko: *Atlantic Hydrophysical Polygon-70*
12. M.A. Mardzhanishvili: *Seismic Design of Frame-panel Buildings and Their Structural Members*
13. E'.A. Sokolenko (editor): *Water and Salt Regimes of Soils: Modeling and Management*
14. A.P. Bocharov: *A Description of Devices Used in the Study of Wind Erosion of Soils*
15. E.S. Artsybashev: *Forest Fires and Their Control*
16. R.Kh. Makasheva: *The Pea*
17. N.G. Kondrashova: *Shipboard Refrigeration and Fish Processing Equipment*
18. S.M. Uspenskii: *Life in High Latitudes*
19. A.V. Rozova: *Biostratigraphic Zoning and Trilobites of the Upper Cambrian and Lower Ordovician of the Northwestern Siberian Platform*
20. N.I. Barkov: *Ice Shelves of Antarctica*
21. V.P. Averkiev: *Shipboard Fish Scouting and Electronavigational Equipment*
22. D.F. Petrov (Editor-in-Chief): *Apomixis and Its Role in Evolution and Breeding*
23. G.A. Mchedlidze: *General Features of the Paleobiological Evolution of Cetacea*
24. M.G. Ravich et al.: *Geological Structure of Mac. Robertson Land (East Antarctica)*
25. L.A. Timokhov (editor): *Dynamics of Ice Cover*
26. K.Ya. Kondrat'ev: *Changes in Global Climate*
27. P.S. Nartov: *Disk Soil-Working Implements*
28. V.L. Kontrimavichus (Editor-in-Chief): *Beringia in the Cenozoic Era*
29. S.V. Nerpin & A.F. Chudnovskii: *Heat and Mass Transfer in the Plant–Soil–Air System*
30. T.V. Alekseeva et al.: *Highway Machines*
31. N.I. Klenin et al.: *Agricultural Machines*
32. V.K. Rudnev: *Digging of Soils by Earthmovers with Powered Parts*
33. A.N. Zelenin et al.: *Machines for Moving the Earth*
34. *Systematics, Breeding and Seed Production of Potatoes*
35. D.S. Orlov: *Humus Acids of Soils*
36. M.M. Severnev (editor): *Wear of Agricultural Machine Parts*
37. Kh.A. Khachatryan: *Operation of Soil-working Implements in Hilly Regions*
38. L.V. Gyachev: *Theory of Surfaces of Plow Bottoms*
39. S.V. Kardashevskii et al.: *Testing of Agricultural Technological Processes*
40. M.A. Sadovskii (editor): *Physics of the Earthquake Focus*
41. I.M. Dolgin: *Climate of Antarctica*
42. V.V. Egorov et al.: *Classification and Diagnostics of Soils of the USSR*
43. V.A. Moshkin: *Castor*
44. E'.I. Sarukhanyan: *Structure and Variability of the Antarctic Circumpolar Current*
45. V.A. Shapa (Chief Editor): *Biological Plant Protection*
46. A.I. Zakharova: *Estimation of Seismicity Parameters Using a Computer*
47. M.A. Mardzhanishvili & L.M. Mardzhanishvili: *Theoretical and Experimental Analysis of Members of Earthquake-proof Frame-panel Buildings*
48. S.G. Shul'man: *Seismic Pressure of Water on Hydraulic Structures*
49. Yu.A. Ibad-zade: *Movement of Sediments in Open Channels*
50. I.S. Popushoi (Chief Editor): *Biological and Chemical Methods of Plant Protection*
51. K.V. Novozhilov (Chief Editor): *Microbiological Methods for Biological Control of Pests of Agricultural Crops*
52. K.I. Rossinskii (editor): *Dynamics and Thermal Regimes of Rivers*
53. K.V. Gnedin: *Operating Conditions and Hydraulics of Horizontal Settling Tanks*
54. G.A. Zakladnoi & V.F. Katanova: *Stored-grain Pests and Their Control*
55. Ts.E. Mirtskhulava: *Reliability of Hydro-reclamation Installations*
56. Ia.S. Ageikin: *Off-the-road Mobility of Automobiles*

(continued)

RUSSIAN TRANSLATIONS SERIES

57. A.A. Kmito & Yu.A. Sklyarov: *Pyrheliometry*
58. N.S. Motsonelidze: *Stability and Seismic Resistance of Buttress Dams*
59. Ia.S. Ageikin: *Off-the-road Wheeled and Combined Traction Devices*
60. Iu.N. Fadeev & K.V. Novozhilov: *Integrated Plant Protection*
61. N.A. Izyumova: *Parasitic Fauna of Reservoir Fishes of the USSR and Its Evolution*
62. O.A. Skarlato (Editor-in-Chief): *Investigation oj Monogeneans in the USSR*
63. A.I. Ivanov: *Alfalfa*
64. Z.S. Bronshtein: *Fresh-water Ostracoda*
65. M.G. Chukhrii: *An Atlas of the Ultrastructure of Viruses of Lepidopteran Pests of Plants*
66. E.A. Bosoi et al.: *Theory, Construction and Calculations of Agricultural Machines*, Vol. 1
67. G.A. Avsyuk (Editor-in-Chief): *Data of Glaciological Studies*
68. G.A. Mchedlidze: *Fossil Cetacea of the Caucasus*
69. A.M. Akramkhodzhaev: *Geology and Exploration of Oil- and Gas-bearing Ancient Deltas*
70. N.M. Berezina & D.A. Kaushanskii: *Presowing Irradiation of Plant Seeds*
71. G.U. Lindberg & Z.V. Krasyukova: *Fishes of the Sea of Japan and the Adjacent Areas of the Sea of Okhotsk and the Yellow Sea*
72. N.I. Plotnikov & I.I. Røginets: *Hydrogeology of Ore Deposits*
73. A.V. Balushkin: *Morphological Bases of the Systematics and Phylogeny of the Nototheniid Fishes*
74. E.Z. Pozin et al.: *Coal Cutting by Winning Machines*
75. S.S. Shul'man: *Myxosporidia of the USSR*
76. G.N. Gogonenkov: *Seismic Prospecting for Sedimentary Formations*
77. I.M. Batugina & I.M. Petukhov: *Geodynamic Zoning of Mineral Deposits for Planning and Exploitation of Mines*
78. I.I. Abramovich & I.G. Klushin: *Geodynamics and Metallogeny of Folded Belts*
79. M.V. Mina: *Microevolution of Fishes*
80. K.V. Konyaev: *Spectral Analysis of Physical Oceanographic Data*
81. A.I. Tseitlin & A.A. Kusainov: *Role of Internal Friction in Dynamic Analysis of Structures*
82. E.A. Kozlov: *Migration in Seismic Prospecting*
83. E.S. Bosoi et al.: *Theory, Construction and Calculations of Agricultural Machines*, Vol. 2
84. B.B. Kudryashov and A.M. Yakovlev: *Drilling in the Permafrost*
85. T.T. Klubova: *Clayey Reservoirs of Oil and Gas*
86. G.I. Amurskii et al.: *Remote-sensing Methods in Studying Tectonic Fractures in Oil- and Gas-bearing Formations*
87. A.V. Razvalyaev: *Continental Rift Formation and Its Prehistory*
88. V.A. Ivovich and L.N. Pokrovskii: *Dynamic Analysis of Suspended Roof Systems*
89. N.P. Kozlov (Technical Editor): *Earth's Nature from Space*
90. M.M. Grachevskii and A.S. Kravchuk: *Hydrocarbon Potential of Oceanic Reefs of the World*
91. K.V. Mikhailov et al.: *Polymer Concretes and Their Structural Uses*
92. D.S. Orlov: *Soil Chemistry*
93. L.S. Belousova & L.V. Denisova: *Rare Plants of the World*
94. T.I. Frolova et al.: *Magmatism and Transformation of Active Areas of the Earth's Crust*
95. Z.G. Ter-Martirosyan: *Rheological Parameters of Soils and Design of Foundations*
96. S.N. Alekseev et al.: *Durability of Reinforced Concrete in Aggressive Media*
97. F.P. Glushikhin et al.: *Modelling in Geomechanics*
98. I.A. Krupenikov: *History of Soil Science*
99. V.A. Burkov: *General Circulation of the World Ocean*
100. B.V. Preobrazhensky: *Contemporary Reefs*
101. V.P. Petrukhin: *Construction of Structures on Saline Soils*
102. V.V. Bulatov: *Geomechanics of Deep-seated Deposits*
103. A.A. Kuzmenko et al.: *Seismic Effects of Blasting in Rock*
104. L.M. Plotnikov: *Shear Structures in Layered Geological Bodies*
105. M.I. Petrosyan: *Rock Breakage by Blasting*
106. O.L. Kuznetsov and E.M. Simkin: *Transformation and Interaction of Geophysical Fields in the Lithosphere*
107. M.V. Abdulov: *Phase Transformations and Rock Genesis*
108. I.E. Mileikovskii and S.I. Trushin: *Analysis of Thin-walled Structures*

(continued on title facing page)

MODELLING THE EFFECTS OF BLASTING ON ROCK BREAKAGE

RUSSIAN TRANSLATIONS SERIES

109. V.Sh. Barbakadze et al.: *Durability of Building Structures and Constructions from Composite Materials*
110. V.I. Kazansky: *Evolution of Ore-bearing Precambrian Structures*
111. D.S. Orlov: *Humic Substances of Soils and General Theory of Humification*
112. V.S. Voitenko: *Applied Geomechanics in Drilling*
113. F.N. Rabinovich: *Concretes with Dispersed Reinforcement*
114. V.A. Borovikov and I.F. Vanyagin: *Modelling the Effects of Blasting on Rock Breakage*

Modelling the Effects of Blasting on Rock Breakage

V.A. Borovikov
and
I.F. Vanyagin

RUSSIAN TRANSLATIONS SERIES
114

CRC Press
Taylor & Francis Group
Boca Raton London New York

CRC Press is an imprint of the
Taylor & Francis Group, an **informa** business

CRC Press
Taylor & Francis Group
6000 Broken Sound Parkway NW, Suite 300
Boca Raton, FL 33487-2742

ISBN-13: 978-9-054-10222-9 (hbk)

Translation of: Modelirovanie deistviya vzriva pri razruzhenii gornikh porod Nedra, Moscow, 1990.

Visit the Taylor & Francis Web site at
http://www.taylorandfrancis.com

and the CRC Press Web site at
http://www.crcpress.com

Preface

The mining industry has witnessed considerable growth in recent years thanks to the widening application of achievements in fundamental and applied sciences. Among the applied sciences, simulation based on the theory of similitude and the dimension theory is of paramount importance as it enables obtaining desired results in a short time period while expending minimal resources.

The physical processes investigated in mining sciences are characterised by enormous complexities vis-à-vis difficulties encountered in theoretical studies as well as direct observations in full-scale (natural) conditions.

Rock breakage, a key process in the winning of minerals, has yet to be adequately studied from a theoretical point of view. Many phenomena accompanying the process are solved on the basis of accumulated experience or qualitative experiments conducted in laboratories as well as under field conditions. This can largely be ascribed to the difficulties of describing mathematically the entourage of phenomena consequent to a blast in rocks, the diversity of the rocks themselves as also their non-homogeneity and jointing.

Even in an approximate approach to the theoretical investigation of rocks, taking them to be ideal solid bodies, mathematical description proves difficult. Actually, the principal mechanical property of solid media—to resist volumetric as well as shear deformation—gives rise to a state of stress that can be considered a tensor of stresses (deformations) characterised by nine components.

Non-homogeneity and jointing of natural rocks markedly complicate a mathematical description of the breakage processes. It is simply not possible to develop a unique mathematical model.

Add to the complexity of mathematical representation the inadequacy of our knowledge concerning the behaviour of natural rocks under the effect of intense dynamic loads generated by a blast and the processes occurring in the explosives per se. Hence many phenomena consequent to a blast in rocks as well as the very process of rock breakage by blasting, are currently being studied by formulating model experiments which employ various methods of simulation, based on the theory of similitude and the dimension theory.

In scientific and technical fields simulation refers to a study of a phenomenon on a model of larger or smaller scale under special, usually laboratory conditions. The objective of simulation is to obtain essential data on the physical phenomena and the parameters of the processes occurring in nature, based on the results of experiments with models. In other words, the phenomenon in which we are interested is duplicated by a physically similar one that can be more conveniently and advantageously studied.

Hence the simulation of phenomena is based on the concept of similitude of phenomena, i.e., the *similitude theory*, which is the basis for modelling. This theory provides the theoretical basis for formulation of model experiments and the basis for simulation methodology. The necessary and sufficient conditions for simulation, i.e., requirements of the model, are formulated on the basis of the similitude theory. It thereby becomes possible to develop the methodology for processing the experimental data obtained and for generalising and extending the obtained results to other full-scale objects or engineering structures and buildings. Thus the significance of the theoretical basis for simulation—the theories of similitude and dimension—for science and technology is quite evident. The theories of similitude and dimension stem from the same prerequisites as laid down for pure theoretical and analytical investigations, i.e., the initial and boundary conditions, parameters and co-ordinates of the process under study or the individual phenomena, are considered in both. These theories serve as a connecting link between theoretical and experimental investigations as well as a bridge between theory and practice.

It should be noted, however, that although the theories of similitude and dimension explore certain links and relationships for a rational look into the processing of experimental data, neither can yield more than what is contained in the equations that describe the phenomenon. In other words, neither is a *universal theory* suitable for every case. They do not yield a complete solution but do allow extension of the experimental data to a field constrained by similitude conditions. The task of formulating an experiment based on the theories of similitude and dimension becomes greatly simplified in such cases, as the functional relationship between entire complexes of parameters determining the phenomenon can be established.

It thus becomes possible in certain cases to avoid studying the effect of each individual parameter on the process under consideration, reducing thereby the number of experiments or greatly simplifying their formulation. Moreover, it is quite possible to extend the experimental results to similar systems within known limits.

While dwelling on simulation, two interrelated aspects have to be kept in view: *model* and *modelling*. By modelling we mean the process, technique or method. The model conceptually reflects a medium through which the phenomenon, process, object under investigation, is subjected to an in-depth study. One distinctive feature of the model is that it does not directly interact with the object under investigation, but replaces it while concomitantly maintaining similarity or conformity with it. Here one can even formulate the so-called *logical* and *imaginary* models, participating in the so-called *vicarious experiments* whose role in understanding any phenomena inaccessible for direct measurements is very significant.

In comparing a model with theory and instrumentation, two conditions are accepted, namely, similarity and representation (replacement). Obviously the absence of similarity would transform a model into a certain symbol or sign. A third condition, featuring the possibility of extrapolation, transforms a model and modelling into a scientific method used conscientiously and in an object-oriented manner in the investigations, as opposed to random observations, inductive generalisations and generally any other spontaneous representation.

It must be borne in mind that a model should not be combined with the usual physical experiment. In the usual experiment the media of experimental studies interact directly with the object under study (natural, full-scale), while in the model experiment interaction is only with the model replacing the full-scale actuality.

In addition to *vicarious experiments*, three other types of simulations are currently used in scientific investigations: physical, mathematical and functional. The first two types have been widely recognised and are dealt with in detail later.

Physical modelling is normally adopted for studying a phenomenon whose mechanism is still not fully known, for example, the process of rock breakage by blasting. In such a case equations interrelating the parameters are derived, the structure of dimensionless governing criteria established and a model prepared for conducting experiments on the basis of general equations describing the process, or from an analysis and consideration of the key parameters and factors characterising the process. In a very simple case the model usually reproduces the phenomenon under study while preserving the geometric similitude and physical nature; it differs from the full-scale by dimensions and speed of occurrence of the process. In some cases it is convenient to fabricate the model from a medium that differs from nature, while compulsorily adhering to the similitude criteria, i.e., simulation in an analogous model or through similar phenomena.

Mathematical modelling is used for all such phenomena as can be mathematically described. Here the investigation is conducted on analogue models which differ in physical composition, but with analogous differential equations that describe the process in the model and the full-scale. Today, analogies between electrical, mechanical, thermal, hydrodynamic, acoustic, diffusive and other physical phenomena are known. Electrical models consisting of active, inductive and capacitance-based resistances and electronic lamps are widely used.

Further growth in technology and science, including theoretical aspects, is inconceivable without experimentation, which means without knowing the simulation technique. The more complex the phenomenon under study, the less we know about it, and the more we resort to trials, to experiments based on the laws of similitude and dimensionality.

The general principles of the theories of similitude and dimension are systematically presented in this monograph and the applicability to phenomena induced by blasting in order to obtain the criteria necessary for similitude and conditions for simulation highlighted. Physical and mathematical modelling of various phenomena and processes induced by blasting in rocks are also considered. These subjects are unfolded in a logical sequence, commencing with elements of the dimension theory and the theorem of similitude and concluding with similitude criteria, conditions for modelling and the requisites of model experiments.

Physical modelling of various phenomena and the process of rock breakage by blasting are dealt with in greater detail. The possibilities of adopting simulation methods developed earlier in mining practice for simulating the process of rock breakage by blasting are also explored. New methods of modelling are suggested by the authors for the study of patterns of formation and propagation of elastic waves in the near-field zone of a blast, stress waves and blast-induced seismic waves in the mid-field and far-field zones of the blast, as well as the processes of expansion of camouflet cavity and quasi-static effect of explosion products generating an additional quasi-static field of stresses, movement of the rock mass and formation of a muck pile and scatter of rock fragments. A new combined method for modelling movement of the rock mass is described, which was successfully adopted in complicated geomining conditions for winning minerals in the direction of a worked-out area filled with water.

The authors have also given a systematic treatment of the existing methods for mathematical modelling and illustrated their application for modelling the laws of stress wave propagation in elastic and elastoplastic media, transmission of these waves through cracks, nature of breakage and fragmentation of rocks by blasting. Also highlighted are the future prospects for applying the methods of mathematical modelling.

Advanced methods for recording high-speed processes that occur in model experiments are likewise detailed in this monograph.

The Introduction and Chapter 3 were written by both authors; Chapters 1, 2 and 4 were written by I.F. Vanyagin.

Contents

1

Theoretical Concepts of Modelling the Effects of Blasting in Different Media

1.1 Elements of Dimensional Analysis

The physical quantities commonly dealt with in scientific investigations and production processes are usually quantified. They are measured in some or the other unit. Further, the quantity is expressed as a numerical value, which characterises the quantity irrespective of whether it has been measured by experimental methods (viz., measurement of length, time, mass) or mathematical expressions (viz., measurement of velocity, acceleration, area, volume etc.).

It is conventional in the dimension theory to classify all physical values into fundamental, derived, non-dimensional or dimensional. The fundamental are those that possess independent dimensions. In other words, the numerical value of a fundamental physical quantity does not change while changing the unit of measurement of another fundamental physical quantity (viz., length, time, mass). Derived physical quantities are those whose numerical values depend on measurement units of fundamental physical quantities (viz., velocity, acceleration, force etc.). Such a distinction between fundamental and derived physical quantities depends entirely on the selection of units of measurement and not on their nature. Hence this classification is a conditional one. In practice, the number of fundamental physical quantities is limited. In mechanics three fundamental quantities are taken into consideration (length, time and mass or force) whereas four fundamental units are considered in electrical technology, heat technology and optics by adding one more to the fundamental three mechanical units, depending on the area of science. For the sake of convenience, the fundamental units have been defined and accepted in all countries and continents, in fact, throughout the whole world. It is known as the system of units for measuring fundamental quantities.

In the former USSR the so-called technical and physical systems of units were long used in mechanics. The fundamental units in technical sciences are length (l), force (F) and time (t), which are measured in metre, kilogramme-force and second respectively (MKS). The fundamental units in physical sciences are length (l), mass (m) and time (t), which are measured in centimetre, gramme and second respectively (CGS).

On 1 January 1968 an international system of units (SI) was introduced by the State Standards. In this system the fundamental unit for length l is metre, for mass m kilogramme, time t second, electric current I ampere, temperature degree Kelvin and intensity of light, candle.

Those quantities whose numerical values depend on the adopted scale and unit of measurement are called dimensional. On the other hand, those quantities whose numerical values do not depend on a unit of measurement are called non-dimensional. Examples of dimensional quantities are length, mass, time, velocity, acceleration, force, energy, moment of a force etc. Examples of non-dimensional quantities are ratio of two lengths, ratio between energy and moment of a force, angles etc.

However, this classification is somewhat conditional. Let us consider the case of an angle. We have labelled it non-dimensional. Nevertheless an angle is measured in units such as radian, degree or percentage of a right angle. Hence the numerical value of an angle depends on a unit of measurement and thus an angle may be termed dimensional. But if an angle is measured as a ratio between the length of an arc and a radius, then whatever the unit of measurement of the values may be, the numerical value of an angle shall be invariant. Hence an angle may be termed non-dimensional. Length can similarly be made non-dimensional by measuring it as a ratio to a fixed unit length. However, it is convenient to measure an angle by a fixed unit since its value does not change with change in scale of the object. But to measure a length the unit of measurement may be varied as the value changes significantly with change in scale. This change may be of the order of hundreds.

The concepts of dimensional and non-dimensional quantities are relative in nature and were formulated by L.I. Sedov: 'These quantities, whose numerical values depend on the selection of a specific unit system from a possible set of unit systems, are called dimensional. Quantities for which the numerical values remain the same in all unit systems of the whole set, are called non-dimensional.'

It may also be noted that the number of fundamental quantities can be either increased or decreased along with the respective increase or decrease in number of dimensional constants or physical constants. An increase in number of fundamental quantities is associated with the introduction of new dimensional physical constants, whereas a decrease in number of fundamental quantities makes physical constants absolute and they become absolute non-dimensional quantities. Hence physical constants may be dimensional or non-dimensional depending on the chosen number of fundamental units of measurement. However, this division is also conditional and not related to the basic nature of these constants.

When the units of measurement of fundamental quantities are fixed, the units of measurement of derived quantities are automatically obtained from their definition. Expressions for units of measurement of derived quantities expressed in terms of units of measurement of fundamental quantities, are called dimensions of these physical quantities. A dimension is symbolically expressed by formulae in which the symbols of the units of length, time, mass, force (in technical unit system) and electric current are expressed by the letters L, T, M, K and I respectively. A symbol $[P]$ is used to denote the dimension of a certain quantity P.

Broadly speaking, dimension may be defined as follows. Let us consider three fundamental quantities A, B, C and a derived quantity P. Let us now change the scale in which the quantities A, B and C are measured, by α, β and γ times. Then the numerical value of the quantity P is changed by π times. In this case, if $\pi = \alpha^k\beta^m\gamma^n$, the quantity P has a dimension

$$[A]^k[B]^m[C]^n, \text{ i.e.,}$$
$$[P] = [A]^k[B]^m[C]^n.$$

This formula shows how the measure of a quantity depends on fundamental quantities A, B and C. This formula is called a dimensional formula. It can be shown that if π is a single-value function of α, β, γ, it must be a higher degree function and the quantity must possess a given dimension, expressed by the formula given above.

It may be noted that the transformation of units may be undertaken correctly in all calculations because of the introduction of a general concept of dimensions. Furthermore, the correctness of a homogeneous physical formula may be verified by this concept of dimensions.

Let us consider as an example the acceleration due to gravity g, for which the dimension is $[g] = LT^{-2}$ and $\pi(\alpha, \beta) = \alpha\beta^{-2}$. In the CGS system, $g = 981$ cm/s^{-2}.

Let us find its value in the MKS system. Here we take $l' = 1$ m $= 100$ cm $= 100\ l$, since its value is changed by $\alpha = 1$ cm/100 cm $= 10^{-2}$ times. The value of t remains unchanged with $\beta = 1$. Then,

$$\pi(\alpha,\beta) = \alpha\beta^{-2} = 10^{-2} \times 1 = 10^{-2}$$

and
$$g' = \pi(\alpha,\beta) \times g = 10^{-2} \times 981 = 9.81 \text{ m/s}^2.$$

In a system, where $l'' = 1$ km $= 10^5$ cm $= 10^5\ l$ and $t'' = 1$ h $= 3600$ s $= 3600\ t$,

$$\alpha = \frac{l}{l''} = \frac{1 \text{ cm}}{10^5 \text{ cm}} = 10^{-5};\ \beta = \frac{1 \text{ s}}{3600 \text{ s}} = \frac{1}{3600},$$
$$\pi(\alpha,\beta) = \alpha\beta^{-2} = 10^{-5}(1/3600)^{-2} = 36^2 \times 10,$$

$$g'' = \pi(\alpha,\beta)g = 36^2 \times 10 \times 981 = 137 \times 10^3 \text{ km/h}^2.$$

Using the above example, it is easy to establish the procedure for changing from one unit system to another. For this, the following steps are required:

1) For the dimension of a quantity P

$$[P] = [A]^k[B]^m[C]^n,$$

the expression

$$\pi(\alpha,\beta,\gamma) = \alpha^k\beta^m\gamma^n$$

should be written.

2) The values of α, β, and γ are determined as ratios of corresponding units of measurement of fundamental quantities in the former system of measurement and the new one expressed in the units of the former system of measurement:

$$\alpha = A/A'; \quad \beta = B/B'; \quad \gamma = C/C'.$$

3) The values of $\pi(\alpha,\beta,\gamma)$ are determined accordingly with the expression

$$\pi(\alpha,\beta,\gamma) = (A/A')^k(B/B')^m(C/C')^n.$$

4) Knowing $\pi(\alpha,\beta,\gamma)$, the value of the derived quantity P' is calculated in the new unit system of measurement by the formula

$$P' = \pi(\alpha,\beta,\gamma)P,$$

where P is the value of the derived quantity in the former unit system of measurement.

1.2 Fundamental Theorems and Additional Conditions of Similarity

1.2.1 Parameters Determining the Class and Group of Phenomena and a Unique Phenomenon

A *class of physical phenomena* is the totality of phenomena described by a common system of fundamental equations (mostly differential equations) with identical physical concepts of the parameters in these equations. Different classes of phenomena are grouped together on the basis of commonality of the fundamental differential equations describing them. A *unique phenomenon* is a member of a class of physical phenomena which differs from the others by numerical values of some given quantities or parameters.

Let us cite as an example of a group of phenomena those described by elliptical differential equations. These equations, in general, may represent the laws of stationary displacement of particles in a medium of a three-dimensional space either in the presence of an external force (non-homogeneous equation, Poisson equation) or in the absence of this (homogeneous equation, Laplace's equation). A class of phenomena may be specified from this group of phenomena. One class, for example, comprises the phenomena of motion of water or air particles, which are characterised by potential speed φ. From this class a unique phenomenon may be identified, which is characterised by discrete single-value conditions.

The following are the *single-value conditions:*

1) Single-value physical conditions are concrete properties of matter given by laws of motion (equations of state of matter or its characteristics).

2) Spatial or geometric single-value conditions are geometric shape and space dimensions in which a phenomenon occurs. These determine the dimensions of the phenomena and also the selection of coordinate systems simultaneous with boundary conditions.

3) Time conditions, at which a given phenomenon starts and occurs, i.e., initial conditions and stationarity or non-stationarity of a process.

4) Boundary conditions for interaction with the surrounding medium, i.e., conditions at boundary of a space occupied by the body in which the phenomenon or process occurs.

An example of a unique phenomenon from a class of phenomena described by hyperbolic equations is the motion of a shock wave in water or the motion of a gas bubble when an explosive is blasted in water.

Every phenomenon of a class or group is characterised by a definite number of fundamental parameters or factors that reflect the essence of processes occurring in the phenomenon.

Hence, one of the primary goals for researchers is to define these parameters and factors. To identify the fundamental parameters which describe a unique phenomenon, a model is developed schematically by the simplest means possible. But it is very difficult to develop a model schematically. A high level of skill, intuition and considerable experience are required for understanding the actual mechanism of the process. Distinguishing the governing parameters and an in-depth study of the patterns and links between them are paramount conditions for the successful solution of a problem.

While defining a phenomenon, it is necessary to identify all its parameters. It is further necessary to know whether these parameters maintain their constant values or not. Thus while considering a physical process in which the weight of a body is involved, we must take acceleration due to gravity (g) into consideration in spite of the fact that it is a constant for most real physical phenomena.

A system of parameters which define a phenomenon can readily be written when the problem is formulated mathematically. In this case all the dimensional and non-dimensional quantities are taken into account, which are necessary and sufficient for determining the required numerical values from the solution of mathematical equations. In all other cases the factors necessary for obtaining the required value are established by intuition, experiments or schematic representation of a phenomenon. In this case the numerical values of parameters can be found only through experiments.

Some quantities with dimension must exist amongst the governing parameters, through which the dimension of such quantities is expressed. Physical constants with dimension may also be included in these parameters.

In many cases the physical quantities which describe a phenomenon are divided into parameters and generalised co-ordinates. Generalised co-ordinates describe the behaviour, motion, change of medium, systems etc. Their number, in general, is equal to the number of degrees of freedom. These are displacements of points, their velocity or acceleration etc.

1.2.2 Concept of Similarity

Let us consider two systems of identical physical nature consisting of an equal number of similar elements, which play an identical role in both systems. By a system we mean here the totality of physical objects or elements which may be grouped together on the basis of some common objective that imparts to the system certain qualities. Examples of such systems are various machines, constructions, magnetic field with its source, stress field in the soil due to blasting, other sources of disturbances etc.

The *parameters of a system* are those quantities which characterise its elements and all external objects interacting with these elements. In other words, these are quantities which isolate the system from the totality of other systems having the same physical nature.

The *common co-ordinates of a medium* are those quantities which describe the behaviour and motion of a system. In general, the common coordinates may change with time and also during transition from one point to another in the system.

Only those systems are considered for which the condition can be described by single-value functions of individual elements. Hence a minimum number of possible parameters which suffice for unique determination of a system and its elements, are separated from the whole set of parameters of a system.

One such system is termed a model and the other one may be termed *au naturel* or actual (full-scale). All quantities related to these model and full-scale systems will henceforth be identified by the indices 'M' and 'N' respectively.

Let a system be defined by m independent parameters P_k consisting of n generalised co-ordinates q_i. The parameters may represent time and conditions of unique definition of the system (initial and boundary conditions, dimensions of the system etc.). Let us consider that the index i has the values 1, 2, ..., n and the index k the values $(n+1)$, ..., $(n+m)$.

The generalised co-ordinates of these systems are similar functions of parameters due to the identical physical nature of the systems. Mathematically, these can be expressed as:

$$q_{M_i} = f_i(P_{M_{(n+1)}}, \ldots, P_{M_{(n+m)}}),$$
$$q_{N_i} = f_i(P_{N_{(n+1)}}, \ldots, P_{N_{(n+m)}}),$$

whereas, in general, $P_{M_k} \neq P_{N_k}$ and $q_{M_i} \neq q_{N_i}$.

Two systems are called similar whenever any two of the generalised co-ordinates for any time and point in the space are proportional, i.e., if:

$$q_{N_i}\,(t_N, r_N) = q_{c_i} \cdot q_{M_i}\,(t_M, r_M),$$
$$r_N = l_c r_M, t_N = t_c t_M,$$

where r_N and r_M are vectors representing points in space; t_N and t_M are time factors; l_c, t_c, q_{c_i} are coefficients of similarity.

If it is known that two systems are similar and it is possible to determine the coefficients of similarity, then it is possible to ascertain the behaviour of one (full-scale) by assessing the behaviour of the other (model).

The simplest example of similarity of systems is geometric similarity (spatial similarity), which is expressed by equalising all corresponding angles and corresponding proportionality of all sides. In other words, this follows the following relationships:

$$\varphi_{N_i} = \varphi_{M_i}, \; x_{N_i} = x_c x_{M_i} = l_c x_{M_i},$$

where $x_c = l_c$ is the scale (coefficient) of geometric similarity; $\varphi_{N_i}, \varphi_{M_i}$ are angles in the original system and the model respectively; x_{N_i}, x_{M_i} the linear dimensions (spatial co-ordinates) in the original system and model respectively.

It may be noted that geometric similarity is a special case of physical similarity and any physical similarity tends to geometric similarity of its respective physical parameters.

Examples of simple similarity are time (synchronous phenomenon or process), kinematic (geometric similarity of velocity and acceleration fields) and dynamic (geometric similarity of force fields).

Time similarity is expressed by the proportionality of time intervals between respective moments of time, i.e.,

$$t_{N_i} = t_c t_{M_i}.$$

Kinematic similarity is expressed by respective directions and proportionality of all velocities and accelerations, i.e.,

$$(v_{x_i})_N = v_{x_c}(v_{x_i})_M; \ (a_{x_i})_N = a_c(a_{x_i})_M$$

or

$$(v_{x_i})_N = \frac{l_c}{t_c}(v_{x_i})_M; \ (a_{x_i})_N = \frac{l_c}{t_c^2}(a_{x_i})_M. \qquad \ldots (1.1)$$

Dynamic similarity is expressed by directions and proportionality of all acting forces, work done by such forces and their power:

$$F_{N_i} = m_{N_i}a_{N_i} = F_cF_{M_i} = m_ca_ca_{M_i}m_{M_i},$$
$$F_{N_i} = a_cm_cF_{M_i} = (m_cl_c/t_c^2)F_{M_i},$$
$$A_{N_i} = F_{N_i}l_{N_i} = F_cl_cA_{M_i} = (m_cl_c^2/t_c^2)A_{M_i},$$
$$N_{M_i} = A_{N_i}/t_{N_i} = (A_c/t_c)N_{M_i} = (m_cl_c^2/t_c^3)N_{M_i}, \qquad \ldots (1.2)$$

where m_c is the scale of mass (material similarity),

$$F_c = m_cl_c/t_c^2; \ A_c = m_cl_c^2/t_c^2; \ N_c = m_cl_c^2/t_c^3. \qquad \ldots (1.3)$$

The foregoing are the expressions for the scale of similarity of forces, work and power respectively.

From expressions (1.1) to (1.3) it is not difficult to conclude that the similarity coefficients in kinematic and dynamic similarities are expressed through the similarity coefficients of initial fundamental quantities $m_c l_c, t_c$ in correspondence with the dimensional formula of these quantities.

Often the similarity coefficients of one kind of derived quantities are expressed readily by similarity coefficients of other derived quantities. Thus while investigating forces acting on machines operating with different velocity regimes, it is better to express the scale of force $F_c = m_cl_c/t_c^2$ in another form. Assuming that the two machines were constructed of the same material (identical density $\rho_c = \rho_M/\rho_M = 1$), we get $m_c = \rho_cl_c^3 = l_c^3$. Hence we can state

$$F_c = m_cl_c/t_c^2 = l_c^3l_c/t_c^2 = l_c^2l_c^2/t_c^2 = l_c^2v_c^2.$$

If in this case the machines possessed identical dimensions, i.e., $l_c = 1$, then $F_c = v_c^2$.

This implies that when a machine is tested under different working force regimes, the forces acting will change proportional to the square of velocity. This fact should be considered while analysing test results.

1.2.3 Fundamental Theorems and Additional Conditions of Similarity

The *first theorem of similarity* was broadly formulated by Academician M.V. Kirpichev. According to him, if a set of phenomena, defined by systems of differential equations and single-value conditions, forms a group of similar phenomena, then the quantities included in the system of equations and single-value conditions must generate complex conditions having identical numerical values for a similar set of phenomena. These complex conditions are called invariants of similarity or similarity criteria. In other words, the necessary condition of similarity for two systems is the numerical equality of similarity criteria, expressed by generalised co-ordinates and parameters of these systems.

This theorem was first formulated for solid bodies by Newton. It can be proven by the Group Theory of Li.

Let us explain the theorem through the following example. Let there be a system consisting of material points, the masses of which are in constant proportion with corresponding points of another system (m_c). The distances and time are similar in the two systems.

Based on Newton's second law the forces acting on these systems will be equal:

$$(F_i)_1 = (m_i)_1 \left(\frac{dl_i}{dt_i^2}\right)_1; \quad (F_i)_2 = (m_i)_2 \left(\frac{dl_i}{dt_i^2}\right)_2;$$
$$(m_i)_1/(m_i)_2 = m_c; \quad (l_i)_1/(l_i)_2 = l_c; \quad (t_i)_1/(t_i)_2 = t_c.$$

Then $(F_i)_1 = m_c \frac{l_c}{t_c^2}(m_i)_2 \left(\frac{dl_i}{dt_i^2}\right)_2 = \frac{m_c l_c}{t_c^2}(F_i)_2 = F_c(F_i)_2$, i.e., these systems are in dynamic similarity. Hence

$$F_c = \frac{m_c l_c}{t_c^2} \quad \text{or} \quad \frac{m_c l_c}{F_c t_c^2} = 1.$$

Therefore, from the values of F, m, l, t a complex can be formed, the numerical values of which will be identical for both systems, i.e.,

$$\frac{(m_i)_1 (l_i)_1}{(F_i)_1 (t_i^2)_1} = \frac{(m_i)_2 (l_i)_2}{(F_i)_2 (t_i^2)_2} = \frac{ml}{Ft^2} = N.$$

Thus a non-dimensional combination of different values is found for the two similar systems. This expression is called the criterion of dynamic similarity. It is also called Newton's criterion.

The *second theorem of similarity* (π-theorem) is the general dimension theory (dimension method) and is formulated as follows: a functional relationship exists in all phenomena not only between real values, which characterise the given phenomenon, but also between some of their non-dimensional combinations, which are necessary for these phenomena to become similar by the criteria of similarity. Proof of this theorem was advanced by Academician L.I. Sedov. Here only the physical concept of the theorem and a few of its important axioms are given and amply illustrated.

Physical relationships, established either theoretically or experimentally, are the functional relationships between different quantities which characterise the given process or phenomenon. The numerical values of these quantities depend on the selection of units of measurement, which are not physically related to the process or phenomenon. Therefore the functional relationships, independent of the systems of units of measurement, must possess a definite structure.

Let a quantity u be a function of some of the independent variables* x_1, $x_2, \ldots, x_n$. These variables include both dimensional and non-dimensional constants.

This is expressed as:

$$u = f(x_1,\ x_2,\ \ldots,\ x_n). \qquad \ldots (1.4)$$

Let us consider that the quantities $u, x_1, x_2, \ldots, x_n$ have their own dimensions and the identity (1.4) is valid irrespective of the choice of system of units of measurement. Also, let us consider that the identity expresses only one physical relationship, indicating that the quantity u has a definite relationship with mutually independent governing parameters $x_1,\ x_2,\ \ldots,\ x_n$. In other words, this expression is not intended to be a mathematical description of a functional relationship; it merely expresses the following identity:

$$[u] = [f(x_1,\ x_2,\ \ldots,\ x_n)]. \qquad \ldots (1.5)$$

Condition (1.5) considerably narrows down the possible class of functions $f(x_1,\ \ldots,\ x_n)$ and allows us to assume to some extent the structure of this physical functional relationship, as can be illustrated by the following two examples:

1. A point starts moving from a state of rest with acceleration a. The path l traversed by the point will be a function of a and time t, i.e., $l = f(a, t)$.

In accordance with identity (1.5), we get

$$[l] = L = [f(a, t)] = LT^{-2}T.$$

Let us express the function $l = f(a, t)$ in the form of an exponential relationship, $l = Ca^n t^m$.

Considering the last two expressions, one may write:

$$L = (LT^{-2})^n (T)^m = L^n T^{m-2n}$$

Equating the indices for dimensions of L and T in the left and right sides:

$$\left.\begin{aligned} 1 &= n, \\ 0 &= m - 2n \end{aligned}\right\} \quad \text{hence, } \begin{aligned} n &= 1 \\ m &= 2n = 2. \end{aligned}$$

---

* In this case, in general, function (1.4) may also be multivalued and may contain so-called 'parasite' relationships, viz., $F = ma$.

The structure of the function will finally be: $l = Cat^2$, where C is a non-dimensional constant.

2. Let us take the example of a mathematical pendulum. An initial velocity v_0 is given to the pendulum and as a result it is deflected by an angle φ. We now establish the relationship of the angle φ with the other parameters that determine the motion of the pendulum. It is obvious that in addition to v_0, we require to know the length of the pendulum l and acceleration due to gravity g, i.e.,

$$\varphi = f(v_0, l, g).$$

In a similar manner, we obtain

$$[\varphi] = 1 = [f(v_0, l, g)], \qquad \ldots (1.6)$$

$$f(v_0, l, g) = Cv_0^k l^n g^m \qquad \ldots (1.7)$$

or

$$1 = L^0 T^0 = (LT^{-1})^k L^n (LT^{-2})^m = L^{k+n+m} T^{-k-2m}$$

Hence,

$$\left.\begin{aligned} 0 &= k + n + m, \\ 0 &= k - 2m. \end{aligned}\right\}$$

Now we have two equations with three unknown variables. This means the problem has no unique solution. Relationship (1.6) may satisfy numerous exponential non-dimensional combinations, one of which may be $v_0^2/(gl)$. The other combinations will be of the nature:

$$[v_0^2/gl]^p,$$

where p is any number.

In this case the nature of function (1.7) could be exponential:

$$\varphi = f(v_0, l, g) = C[v_0^2/(gl)]^p,$$

or may be more complex:

$$\varphi = f(v_0, l, g) = v_0^2/(gl) + 2(v_0^2/(gl))^2 - 4(v_0^2/(gl))^4,$$

$$\varphi = \sin(v_0^2/(gl))^2 \quad \text{or} \quad \varphi = \log(v_0^2/(gl)).$$

It is clear from these examples that dimensional analysis gives no general mathematical solution for a phenomenon. However, narrow may be the range of possible functional relationships, dimensional analysis gives only the assumed structure of the relationship.

Let us go back to the general case of functional relationship (1.4). Here we shall see, the limitations imposed by relationship (1.5) on function (1.4). Also, we shall explain the structure of the function, which expresses a definite physical law, independent of the choice of system of units of measurement.

For this, k fundamental quantities $x_1, x_2, \ldots, x_k$ are arbitrarily chosen from the totality of $n + 1$ independent quantities or parameters in the functional relationship expressed in a generalised form, whereas complexes of the following types are constructed by following the rules of formulae of dimension from the remaining quantities $x_{k+1}, \ldots, x_n, u$:

$$\frac{u}{x_1^{n_1} x_2^{n_2} \ldots x_k^{n_k}}; \quad \frac{x_{k+1}}{x_1^{m_1} x_2^{m_2} \ldots x_k^{m_k}}; \quad \frac{x_n}{x_1^{q_1} x_2^{q_2} \ldots x_k^{q_k}},$$

where the denominators show the dimensions of respective quantities given in the numerators. As a result, functional relationship (1.4) may be expressed as follows:

$$\Pi = f(\Pi_1, \Pi_2, \ldots, \Pi_{n-k}), \qquad \ldots (1.8)$$

where

$$\Pi = \frac{u}{x_1^{n_1} x_2^{n_2} \ldots x_k^{n_k}}, \quad \Pi_1 = \frac{x_{k+1}}{x_1^{m_1} \ldots x_k^{m_k}},$$

$$\Pi_{n-k} = \frac{x_n}{x_1^{q_1} x_2^{q_2} \ldots x_k^{q_k}} \qquad \ldots (1.9)$$

The above expressions (1.9) are non-dimensional similarity criteria.

Two different situations may arise in this case:

1) The quantities $x_1, x_2, \ldots, x_n$ may give rise to non-dimensional exponential combinations of type (1.9). In that case the following type of relationship is obtained:

$$u = f_1(x_1, x_2, \ldots, x_k) \cdot f_2(\Pi_1, \Pi_2, \ldots, \Pi_{n-k}), \qquad \ldots (1.10)$$

where $f_1(x_1, x_2, \ldots, x_k) = x_1^{n_1} x_2^{n_2} \ldots x_k^{n_k}$.

2) Non-dimensional exponential combinations (1.9) are impossible. In such a case the relationship of u will be

$$u = Cx_1^{n_1} x_2^{n_2} \ldots x_k^{n_k}, \qquad \ldots (1.11)$$

where C is some arbitrary constant.

Formula (1.11), as expressed in the second case, gives a complete idea about the character of the relationship being sought.

It is also to be noted that the smaller the number of quantities to be determined, the more constrained the functional relationship; in such cases the investigation becomes easier. When $n \leqslant k$, as shown by formula (1.11), the dimension method determines the functional relationship with an accuracy up to the value of the arbitrary constant. This is the most convenient situation. But an artificial increase in the number k does not help in this situation, since the number of parameters would concomitantly increase in the form of dimensional constants while the number of non-dimensional combinations would remain unchanged. In some cases a few similarity criteria may be grouped together by either multiplication or division. This may reduce their numbers. However, such new criteria, obtained in this manner, usually maintain all earlier properties. They also simplify the process of obtaining criteria relationships by experiments.

Let us illustrate the situation when $n \leqslant k$ by simple examples taken from special cases of modelling.

By special cases of modelling we mean modelling of processes which take place in a unique field of force (viz., either forces of inertia or forces of viscosity). In such cases not more than one criterion may be formulated.

Such cases of modelling are often found in practice. These are observed when a phenomenon is investigated in conditions wherein a limited number of values predominantly define a process. In such a case the characteristics of the process taking place in a field of forces determine such forces and the quantities that characterise the object on which these forces act. From these forces the unique similarity criterion is formulated.

Example 1. Resistance to motion of a solid body by water and air.

A) Let us first consider an ideal case wherein motion is defined by forces of inertia alone, $F = ma$. Hence, expressing the right side as

$$ma = \rho l^3 a = \rho l^3 l/t^2 = \rho l^2 v^2, \text{ we get}$$

$$F = ma = \rho l^2 v^2. \qquad \ldots (1.12)$$

From this identity, considering the main parameters as ρ, l, v, we may get the non-dimensional similarity criterion by dividing one part of the identity by another, i.e., we may get Newton's similarity criterion:

$$F/(ma) = F/(\rho l^2 v^2) = N. \qquad \ldots (1.13)$$

From expression (1.13), Newton's formula is easily obtained for forces of resistance to the motion of solid bodies in a liquid with an accuracy up to a certain constant, C, that depends on the shape of a similar geometric body:

$$F = C\rho D^2 v^2,$$

where D is some characteristic dimension of the body. Considering expression (1.13) and the relationship for an aerodynamic force, we get

$$F = R = C_R \frac{\rho v^2}{2} S,$$

where C_R is the resistance coefficient and equal to $2C$;
S is the area of middle section, $S \approx l^2$

B) Let us now consider that the liquid is not an ideal one, but rather a viscous one. Here the governing force will be viscosity:

$$F = \eta S d\varphi/dt,$$

where φ is the shear angle of layers of the liquid;
η is the dynamic coefficient of viscosity;
S is the area of cross-section of the plate (body) affected by the liquid during motion.

This expression can be written by an analogy with the previous example as:

$$F = \eta l^2/t.$$

Equating this relationship with (1.12), we get:

$$\eta l^2/t = \rho l^2 v^2,$$
$$\rho v^2 t/\eta = \rho v l/\eta = v l/(\eta/\rho) = v l/\xi.$$

A new non-dimensional combination is thus discovered, the similarity criterion, which is called the Reynolds number, i.e.,

$$Re = v l/\xi, \qquad \ldots (1.14)$$

where $\xi = \eta/\rho$ is the kinematic coefficient of viscosity.

The Reynolds number characterises the relationship between the forces of viscosity and inertial forces. The character of flow of a liquid changes (laminar or turbulent) according to the Reynolds number.

Modelling in the same medium as encountered in reality and considering the Reynolds criterion, gives these similarity scales:

$$v_c = v_N/v_M; \; l_c = l_N/l_M; \; \rho_c = \rho_N/\rho_M;$$
$$\xi_c = \xi_N/\xi_M; \; \rho_c = \eta_c = 1; \; v_c l_c = 1; \; t_c = l_c^2.$$

This means that the time scale will be proportional to the square of the linear scale, i.e., processes in the model must occur notably faster than in nature by approximately l_c^2 times.

Example 2. Major force, the force of elasticity

A) Let us first observe the motion of a weight attached to a spring. The force of elongation of the spring is $F = Kx$, where K is a coefficient of proportionality, N/m and x the linear dimension of the spring, m.

Equating this force to the force of inertia, we get

$$Kx = ml/t^2.$$

From this a non-dimensional combination may be formed by dividing one part by the other, which is $m/(kt^2)$. An expression for time t may also be obtained:

$$t^2 = m/K \text{ or } t = \sqrt{m/K}.$$

From this expression the relationship for the period of oscillation of the spring is readily obtained:

$$T = C\sqrt{m/k} \text{ or } T = 2\pi\sqrt{m/K},$$

where the value of $C = 2\pi$ may be established by experiments.

B) Let us now consider elastic vibrations in a solid body (stress wave propagation in solid bodies). Following Hooke's law, the stress $\sigma = \varepsilon E$, where ε is the strain and E the elastic modulus.

The force of elasticity is $F = \sigma S = \varepsilon ES$, where S is the area of cross-section perpendicular to the direction of the acting force.

By equating the acting force to the inertial (1.12), we get:

$$\varepsilon ES = \rho v^2 l^2,$$

leading into the following non-dimensional combination,

$$\rho v^2/(\varepsilon E) = v^2/(E\varepsilon/\rho) = \text{ idem}$$

and then the similarity criterion of Cauchy after removal of the root in the previous identity:

$$Ca = v/\sqrt{E\varepsilon/\rho}. \qquad \ldots (1.15)$$

This characterises the relationship between elastic forces and the forces of inertia.

It is to be noted that in modelling a process in the same medium, we get: $\rho_c = E_c = \varepsilon_c = 1$, from which $v_c = l_c/t_c = 1$ and $l_c = t_c$, i.e., simple geometric similarity holds good for the processes of elastic wave propagation in similar media.

Example 3. Modelling under gravitational forces:

$$F = \gamma m_1 m_2/r^2, \qquad \ldots (1.16)$$

where γ is the gravitational constant;
m_1 and m_2 the masses of interacting bodies;
r the distance between the bodies.

For the solar system mass m_2 consists of m_1 (mass of individual planets). So,

$$m_1\gamma = K = \text{ const.}$$

For interaction of a planet with the Sun, the following expression may be written using (1.16):

$$F = Km_i/r_i^2.$$

Comparing the above expression with (1.12), we get

$$Km_i/r_i^2 = m_i l_i/t_i^2,$$

whence $t_i^2 = l_i^3/K$. Using this, the law of planetary motion can be given as:

$$t_1^2/t_2^2 = a_1^3/a_2^3, \qquad \ldots (1.17)$$

where a_1 and a_2 are the semi-axes of the elliptical orbit of planetary motion around the Sun.

In this case expression (1.17) describes the third law of Kepler.

Example 4. Modelling the force of gravity near the surface of the Earth. The force of gravity $F = mg$. Comparing this with expression (1.12), we get:

$$mg = ma, g = a = l/t^2,$$

from which two expressions may be obtained:

$$t = \sqrt{l/g} \text{ and } gl = v^2. \qquad \ldots (1.18)$$

Expression (1.18) leads to the famous similarity criterion of Froude:

$$Fr = v^2/(gl). \qquad \ldots (1.19)$$

This characterises the relationship between the force of inertia and the force of gravity.

Example 5. Modelling of centrifugal forces.

The centrifugal force $F = \omega^2 mR$, where $\omega^2 R$ is the centrifugal acceleration; ω angular velocity; R radius of rotation.

Equating this force with the force of gravity $F = mg$, we get $m\omega^2 R = mg$ or $\omega^2 R = g$, from which we get the similarity criterion for centrifugal forces:

$$\omega^2 R/g = \text{idem},$$

which is used for centrifugal modelling.

This similarity criterion is widely used in investigating the process of rock breaking by blasting in a centrifuge when $l_c \neq 1$, i.e., when the dimensions of the model and the actual dimensions do not coincide.

The examples of modelling given above are often encountered in modelling of physical processes. Hence, it is better to have a table of time-space scaling for these special cases of modelling with the condition that the material used in the model and the material in nature be identical, i.e., when $\rho_c = 1$ (Table 1).

To conclude, let us consider some examples that are more complex for application of the theory of dimension ($k < n$).

1. The path traversed by a point. Initially, the point traverses with uniform acceleration with an initial velocity v_0. Thus, the path will be:

$$l = f(v_0, a, t). \qquad \ldots (1.20)$$

Of the three variables (v_0, a, t), two—v_0 and t—are assumed to be independent while the third, a is considered a dependent one. To determine the structure of the functional relationship of equation (1.20), the method of dimensions is applied.

Let us consider the unit system M, L, T. The dimension of all values in this system will be:

$$[l] = L;\ [v_0] = LT^{-1};\ [a] = LT^{-2};\ [t] = T.$$

Table 1

Nomenclature of characteristics	Symbol	Major forces				
		Viscosity $F = Sd\varphi/dt$	Elasticity $F = Kx$	Gravity $F = mg$	Gravitational $F = \gamma m_1 m_2/r^2$	Centrifugal $F = m\omega^2/R$
Indicator of similarity (characteristic number)	Λ	$\frac{v_c l_c}{\xi_c} = 1$	$\frac{v_c}{\frac{\varepsilon_c}{\rho} E_c} = 1$	$v_c^2/(l_c g_c) = 1$	$\frac{l_c v_c^2}{\gamma_c m_c} = 1$	$g_c/(\omega_c^2 R_c) = 1$
Similarity criterion	Π	$Re = vl/\xi$	$Ca = v/\sqrt{\varepsilon E/\rho}$	$Fr = v^2/(lg)$	$lv^2/\gamma m = \text{idem}$	$g/(\omega^2 R) = \text{idem}$
Relationship between time scale and linear scale l_c	t_c	l_c^2	l_c	$l_c^{1/2}$	$l_c^{3/2}$	l_c^{κ}
Relationship between velocity scale and linear scale l_c	$v_c = l_c/t_c$	l_c^{-1}	1	$l_c^{1/2}$	$l_c^{-1/2}$	$l_c^{1-\kappa}$
Relationship between force scale and linear scale l_c at $\rho_{c_*} = 1$	$F_c = l_c^2 v_c^2$	1	l_c^2	l_c^3	l_c	$l_c^{4-2\kappa}$

Contd...

Table 1—Continued

Nomenclature of characteristics	Symbol	Major forces				
		Viscosity $F = Sd\varphi/dt$	Elasticity $F = Kx$	Gravity $F = mg$	Gravitational $F = \gamma m_1 m_2/r^2$	Centrifugal $F = m\omega^2/R$
Relationship between stress scale and linear scale l_c at $\rho_c = 1$	$\sigma_c = v_c^2$	l_c^{-2}	1	l_c	l_c^{-1}	$l_c^{2(1-\kappa)}$
Relationship between work scale and linear scale l_c at $\rho_c = 1$	$A_c = l_c^3 v_c^3$	l_c	l_c^3	l_c^4	l_c^2	$l_c^{5-2\kappa}$
Relationship between power scale and linear scale l_c at $\rho_c = 1$	$N_c = l_c^2 v_c^3$	l_c^{-1}	l_c^2	$l_c^{7/2}$	$l_c^{1/2}$	$l_c^{5-3\kappa}$

Remarks: For centrifugal modelling, the value of κ changes: $\kappa = 1$ for elastic medium, $1 < \kappa < 2$ for elastoviscous medium.

Let us form non-dimensional combinations from fundamental and derived values:

$$\Pi = l/(v_0^x t^y); \; \Pi_1 = a/(v_0^x t^y).$$

From the non-dimensional conditions of Π and Π_1, we obtain:

$$[l] = [v_0^x t^y], \; [a] = [v_0^x t^y]. \qquad \ldots (1.21)$$

For the present unit system of measurement identity (1.21) may be rewritten as:

$$L = (LT^{-1})^x T^y = L^x T^{y-x}; \; LT^{-2} = L^x T^{y-x}.$$

Equating indices L and T, we get the following equations: from the first relationship,

$$\left.\begin{aligned} 1 &= x, \\ 0 &= y - x, \end{aligned}\right\} \quad \begin{aligned} x &= 1 \\ y &= x = 1, \end{aligned} \quad \Pi = \frac{l}{v_0 t};$$

and from the second relationship,

$$\left.\begin{aligned} 1 &= x, \\ -2 &= y - x, \end{aligned}\right\} \quad \begin{aligned} x &= 1, \\ y &= x - 2 = -1, \end{aligned} \quad \Pi_1 = a/(v_0 t^{-1}) = at/v_0.$$

Then the relationship for l may be written as:

$$\Pi = f(\Pi_1) \text{ or } l/(v_0 t) = f(at/v_0)$$

or

$$l = v_0 t f(at/v_0). \qquad \ldots (1.22)$$

The functional relationship established in (1.22) does not give an accurate mathematical expression for the relationship, which is obtained by solving the problem and expressed as:

$$1 = v_0 t + \frac{at^2}{2} = v_0 t \left(1 + \frac{1}{2}\frac{at}{v_0}\right). \qquad \ldots (1.23)$$

This gives an approximate relationship and combination of parameters [see right side of the derived equation (1.23)]. Here the relationship is defined by a combination of parameters $v_0 t$ and at/v_0.

If an experiment is conducted and the data obtained analysed with respect to Π and Π_1, a relationship will readily be obtained as given below:

$$\Pi = (1 + \Pi_1/2).$$

Equation (1.23) may be obtained from this relationship.

This is one of the simplest examples for establishing a functional relationship by the dimension method. Let us consider another example.

2. Let an initial velocity v_0 be applied to a mathematical pendulum that is deflected by an angle φ. We are to find the relationship between the angle

of deflection φ and the parameters that define the motion of the pendulum. The following function is obvious:

$$\varphi = f(v_0, l, g). \qquad \ldots (1.24)$$

Let us consider v_0 and l (may be g and l) to be independent parameters. The system of units is *MLT*. The non-dimensional combination of the parameters are:

$$\Pi = \varphi/(v_0^x l^y), \ \Pi_1 = g/(v_0^x l^y).$$

The dimensions of the variables in (1.24) are:

$$[\varphi] = 1; \ [v_0] = LT^{-1}; \ [l] = L; \ [g] = LT^{-2}.$$

From Π and Π_1, the following identity may be written:

$$[\varphi] = [v_0^x l^y]; \ 1 = L^{x+y}T^{-x};$$
$$[g] = [v_0^x l^y]; \ LT^{-2} = L^{x+y}T^{-x}.$$

From these identities, we get the necessary equations for the indices x and y and also the following non-dimensional parameters Π and Π_1:

$$\text{for } \Pi \ \left.\begin{array}{l} 0 = x + y, \\ 0 = -x, \end{array}\right\} \quad x = y = 0 \ \text{ and } \ \Pi = \varphi;$$

$$\text{for } \Pi_1 \ \left.\begin{array}{r} 1 = x + y, \\ -2 = -x, \end{array}\right\} \begin{array}{l} x = 2; \\ y = 1 - x = -1; \end{array} \quad \Pi_1 = gl/v_0^2.$$

Finally, we derive:

$$\varphi = f(gl/v_0^2). \qquad \ldots (1.25)$$

From the above formula, it can be seen that the variable φ depends on a non-dimensional combination $\Pi_1 = gl/v_0^2$. It may be an exponential function:

$$\varphi = C(gl/v_0^2)^n,$$

where n is a numerical index.

Function (1.25) may be expressed by other more complex functions. They are:

$$\varphi = gl/v_0^2 + 2(gl/v_0^2)^2 + \ldots$$

or $\quad \varphi = \sin(gl/v_0^2); \ \varphi = \ln(gl/v_0^2)$ and so on.

Many possible functions for φ are obtained because φ itself is non-dimensional and so is the combination of governing indices. Thus the dimension method does not give a reasonable answer. In other words, in such cases the problem is more indefinite than in the case presented in example no. 1.

The *third theorem of similarity* establishes the conditions of similarity sufficient for a given set of phenomena. Professor V.L. Kirpichev was the first to prove this theorem, formulated as follows: Two bodies prepared from the same material, which were similar before applying an external force, remain similar after application of the force if the forces are distributed over their surfaces in a similar manner and the values of forces per unit area in both bodies are identical. In other words, according to the generalised formulation of the third theorem of similarity, similarity in the boundary conditions of the uniqueness of phenomena needs to be satisfied. The theorem confirms that the sufficient condition of similarity between two systems is the equality of similarity criterion formed by their parameters and that these parameters remain in the boundary conditions of uniqueness of phenomena.

The proof of the theorem is complex and based on concepts of continuous groups of transformation. Hence it is not to be adduced here.

In addition to these three theorems of similarity, *four supplementary clauses of similarity* between systems play a vital role, namely:

1. Complex systems comprising a number of individual systems are similar both individually and collectively if all elements of the individual systems are similar, i.e., if the conditions of similarity prevail in the elementary relations of these systems.

2. Conditions of similarity which are valid for linear systems may be transferred to non-linear systems with an additional condition involving the coincidence of relative characteristics of non-linearity, i.e., in coincidence of corresponding relationships between variables. In other words, the similarity conditions which are true for systems with fixed parameters, may be applicable for systems with variable parameters whenever the relative characteristics of the latter coincide.

For example, in designing an electromagnet the following conditions are imposed on the magnetic permeability η of the model and the new specimen produced:

$$\frac{\eta_N(H_N)}{\eta_M(H_M)} = \eta_c = \text{ const for } H_N/H_M = H_c = \text{ const},$$

where H is the intensity of the magnetic field.

3. Similarity conditions that are true for isotropic and homogeneous systems, may be applicable in anisotropic and heterogeneous systems, if the anisotropy and heterogeneity in both systems are identical.

4. Processes which take place in geometrically non-similar systems may be similar if every point in the 3-D space of one system corresponds to every well-defined point of the other similar physical system. Such similarity may be termed Athenian.

1.3 Concept of Imperfect and Approximate Similarity

The following difficulties are encountered in solving a concrete problem on similarity of phenomena in nature and a model:

1. Not all the governing parameters of the given phenomenon are known.
2. Of the governing parameters, some may exert considerable influence on the phenomenon under investigation and some very little.
3. It is impossible to sort out the parameters of a natural phenomenon in such a way that the governing criteria of the natural phenomenon are exactly identical with the model.
4. The presence of variable parameters and anisotropy does not satisfy the supplementary similarity conditions.

Other difficulties, not foreseeable in those outlined above, may also arise.

Under these circumstances, imperfect or approximate similarity may be invoked, wherein a phenomenon is described by the limited number of known parameters that influence it most. Those similarity criteria which cannot satisfy both the natural phenomenon and the model simultaneously are disregarded. Here an average of the variable parameters and anisotropic (or non-homogeneous) conditions is taken and only a few of the similarity criteria satisfied.

The degree of approximation of similarity is characterised by the so-called **scale effect**, which is usually quantified by results of studies on how significant is the allowable ceiling on the number of parameters; to what extent the similarity criteria can be neglected and how important is the assumed averaging of the processes under study. The scale effect is evaluated in most cases by special experimental investigations.

It is to be noted that even approximate and partial modelling are also quite valuable for further studies of a phenomenon as they enable definition in certain cases of the principal (key) laws and depiction of the phenomena and processes being studied. Moreover, precise modelling while maintaining complete similarity in the model is not feasible in practice.

The similarity established vis-à-vis to a particular part (not all) of the quantities that characterise the phenomenon under study is termed **partial similarity.** Such a similarity is possible in principle because in most cases researchers are not interested in the entire physical pattern, but only a part of it.

For example, one may seek to determine either reactions at (load) bearing points or internal stresses and deformations.

In the case of partial similarity, it is essential to determine the unique quantities that will influence the results. To do so, naturally one must understand well the physical essence of the problem.

The similarity in which a set of factors exerting insignificant influence on the phenomenon under study are deliberately omitted (in contrast to the partial similarity), is termed **approximate similarity.** Here, it is essential to properly assess the influence of omitted factors on the phenomenon. Errors arising from approximation have also to be evaluated; in other words, the scale effect has to be taken into account.

The physical aspect of partial and approximate similarity is best illustrated through examples.

Example 1. **Partial similarity**: A system of brackets 1, 2, 3 is hinged at A_N on which a force P_N acts (Fig. 1, a). The longitudinal forces in the bars are to be determined.

This hinged-bar construction cannot be determined statically since the longitudinal forces in the bars depend not only on the force P_N, angles α_N and β_N, but also on the rigidity of the bars against longitudinal deformation $i = ES/l$, where E is the modulus of longitudinal elasticity; S is the cross-sectional area of the bar and l is the length of bar.

Thus the quantities that unambiguously define the problem would be: α, β, P, i_1, i_2, i_3 with dimensions $[\alpha] = [\beta] = 1$;

$$[P] = MLT^{-2}; \ [i_1] = [i_2] = [i_3] = MT^{-2}.$$

The angles $[\alpha]$ and $[\beta]$ are non-dimensional and of the four remaining quantities, only two non-dimensional combinations can be formulated, in particular the two simplexes, $i_1/i_2 =$ idem, $i_3/i_2 =$ idem.

The dimensions of P and i are independent; hence P finds no place in any of the similarity criteria and its scale may be arbitrarily assumed as P_c. The scale of extent of rigidity should also be arbitrary, but should be the same for all the bars.

$$(i_1)_c = (i_2)_c = (i_3)_c = i_c. \qquad \ldots (1.26)$$

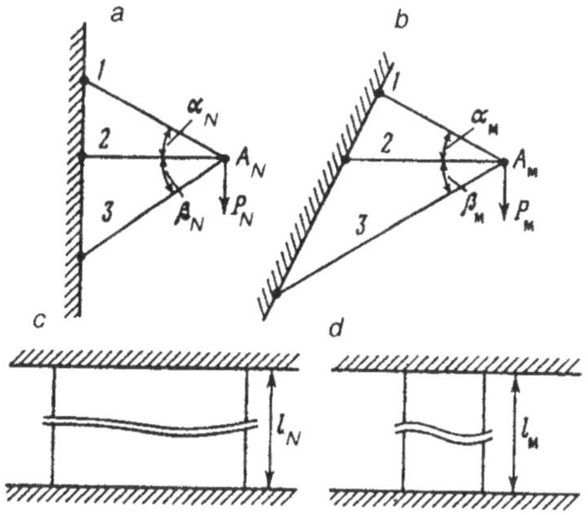

Fig. 1. Diagrams of incomplete modelling of a system of brackets (a—actual, b—model) and approximate modelling of a restrained bar subjected to heating (c—actual, d—model).

It appears, therefore, that the model (Fig. 1, b) may be prepared from different materials with differing values of moduli of elasticity E and Poisson ratio ν. Lengths of bars in the model need not be compulsorily proportional to the actual lengths of bars. It is essential that (1.26) be fulfilled. Then the forces in bars are determined as follows:

$$(N_i)_N = (N_i)_M N_c = (N_i)_M P_c, \; i = 1, \; 2, \; 3.$$

It should be remembered that while modelling one should never transgress the elastic limits within which the material can function.

Example 2. Let us take the problem of a restrained bar (Fig. 1,c) and, in particular, determine the normal stresses in cross-sections of the bar heated uniformly to a temperature of $\Delta\theta$. Let us deliberately ignore the low weight of the bar, i.e., we shall consider approximate modelling.

The governing quantities of the process (from the condition of unambiguity) would be:

E—modulus of elasticity, $[E] = ML^{-1}T^{-2}$;

α—coefficient of linear expansion, $[\alpha] = \theta^{-1}$;

and $\Delta\theta$—temperature, $[\Delta\theta] = \theta$.

From these quantities, only one similarity criterion or non-dimensional combination of governing quantities $\pi = \alpha\Delta\theta$ can be obtained, which corresponds to the relationship between scale parameters $\alpha_c\Delta\theta_c = 1$.

This allows one to conclude that any material with any value of E may be selected for a model; it need not necessarily conform to geometric similarity (Fig. 1,d). The model should have a uniform cross-section over the entire bar length and kinematically identical supports. Then the stresses would be modelled in the following scales:

$$\sigma_c = E_c = E_N/E_M; \; \sigma_N = \sigma_c\sigma_M.$$

In doing so evidently errors would depend on the ratio of lengths and weights of bars in the actual device and in the model, i.e., on

$$l_c = l_N/l_M \text{ and } \gamma_c = \gamma_N/\gamma_M,$$

where l is length of the bar; γ its density.

Let us quantify the error. By accurately solving the problem, the stress in the lower cross-section of the bar is found to be

$$\sigma'_N = \alpha_N E_N \Delta\theta_N + \gamma_N l_N/2. \qquad \ldots (1.27)$$

On the other hand, as $\sigma_N = \sigma_c\sigma_M$ equality (1.27) can be rewritten as:

$$\sigma_N = \sigma_c \sigma_M = \sigma_c(\alpha_M E_M \Delta\theta_M + 0.5\gamma_M l_M) = \sigma_c \left(\frac{\alpha_N E_N \Delta\theta_N}{\alpha_c E_c \Delta\theta_c} + \frac{\gamma_N l_N}{2\gamma_c l_c} \right) =$$

$$\frac{\alpha_N E_N \Delta\theta_N}{\alpha_c \Delta\theta_c} + \frac{E_c}{\gamma_c l_c}(\gamma_N l_N) = \alpha_N E_N \Delta\theta_N + \frac{E_c}{\gamma_c l_c}\left(\frac{\gamma_N l_N}{2}\right),$$

as $\sigma_c = E_c$ and $\alpha_c \Delta\theta_c = 1$.

Error in the computation of stress is given by

$$\Delta\sigma_N = \frac{\sigma_N - \sigma'_N}{\sigma'_N} = \frac{0.5\gamma_N l_N \left(\dfrac{E_c}{\gamma_c l_N} - 1 \right)}{\alpha_N E_N \Delta\theta_N + \dfrac{\gamma_N l_N}{2}} =$$

$$\frac{0.5\gamma_N l_N}{\alpha_N E_N \Delta\theta_N + 0.5\gamma_N l_N} \left(\frac{E_c}{\gamma_c l_c} - 1 \right) = A(\Lambda - 1).$$

Here A denotes the extent of neglecting weight; Λ is the similarity indicator which, for complete similarity, evidently is equal to 1. From this it transpires that the more Λ differs from unity, the greater the error in modelling or the scale effect.

As an example of similarity under dynamic loading, let us consider the general technique of establishing conditions of similarity for a stressed and deformed state under impact.

Let a highly rigid body A weighing P (whose deformation may be neglected), falling from a height H, collide with an immobile fixed construction (system) B.

Displacements $\delta_{D.N}$, stresses $\sigma_{D.N}$ and reaction $R_{D.N}$ in the body subjected to impacts are determined by the known relationships:

$$\delta_{D.N} = K_{D.N}\delta_{st}; \ \sigma_{D.N} = K_{D.N}\sigma_{st}; \ R_{D.N} = K_{D.N}R_{st},$$

where $K_{D.N}$ is the dynamic coefficient; δ_{st}, σ_{st}, R_{st} are static displacements, stresses and reaction respectively.

The dimension of $[K_{D.N}] = 1$. Consequently, for similarity one should consider the similarity of corresponding static quantities while the dynamic coefficient for the model and the actual situation should be the same.

Coefficient $K_{D.N}$ is determined by the various formulae given below:

$$K_{D.N} = 1 + \sqrt{1 + \frac{2H}{\delta_{st}(1 + \beta Q/P)}}; \qquad \ldots (1.28)$$

$$K_{D.N} = 1 + \sqrt{\frac{v^2}{g\delta_{st}(1 + \beta Q/P)}}, \qquad \ldots (1.29)$$

where Q is the weight of the body subjected to impact; β is the dimensionless coefficient that depends on static deformations and weight of the body under impact; v is the velocity of falling body at the time of impact.

It follows from (1.28) that similarity criteria of the following type should be considered for modelling.

$$H/\delta_{st} = \text{idem} \quad \text{or} \quad H_c/(\delta_{st})_c = H_c/l_c = 1,$$

i.e., the height H_c from which the body (making impact) falls should be modelled in a linear scale l_c.

From (1.29) another similarity criterion is evolved.

$$v^2/(g\delta_{st}) = \text{idem} \quad \text{or} \quad v_c^2/[g_c(\delta_{st})_c] = 1.$$

At $g_c = 1$ (which is the usual case)

$$v_c = \sqrt{(\delta_{st})_c} = l_c^{1/2},$$

i.e., the velocity of fall should be proportionate to the square root of linear scale, l_c.

In modelling the mutual collision of solid bodies within elastic limits, as already mentioned, it is necessary under certain conditions of modelling to take into account the similarity criteria of Cauchy.

$$\frac{v}{\sqrt{\varepsilon E/\rho}} = \text{idem} \quad \text{or} \quad \frac{v_e}{\sqrt{\varepsilon_c E_c/\rho_c}} = 1$$

when $E_c = \rho_c = \varepsilon_c = 1$, $v_c = l_c/t_c = 1$, if $l_c = t_c$, i.e., ratio of time periods of mutual impact is proportional to the linear scale.

Finally, it should be noted that all the examples considered above are related to inertial forces or uniformly accelerated processes. Usually under the non-stationary effect of loads the similarity problem would be more complicated and require special investigations in each specific case.

1.4 Fundamental Concepts of Modelling

Modelling of physical phenomena is based on the theory of similitude. It is a scientific method for generalising results of investigations of individual phenomena on other similar phenomena. The phenomena of a particular class having similar characteristic quantities (primarily with similar magnitude of unambiguous conditions) and differing only in the scales of these quantities are termed similar physical phenomena.

The necessary and sufficient conditions for similar physical phenomena are:

1) Invariability (invariance) of system of principal equations when variables are transformed in a similar manner;
2) Similarity of all physical parameters of the system of principal equations;

3) Geometric (spatial) similarity;
4) Stationary nature of the process or similarity of all variables at the initial moment of the process (for non-stationary processes);
5) Similarity of boundary conditions.

All these conditions amount to equality of corresponding points in space and corresponding time periods of specific non-dimensional sets or criteria of similitude. As for similarity of physical phenomena, invariability of only the key similarity criteria suffices, i.e., the similarity criteria formulated from the given magnitudes of conditions of uniqueness (of system parameters). The remaining similarity criteria that are not decisive, i.e., the similarity criteria formulated from the requisite dependent parameters (generalised system of co-ordinates) would automatically coincide.

There is a specific causal relationship between these categories of similarity criteria, as is known from the conclusions of the π-theorem. It is mathematically expressed in the form of a single-value functional relation. This relation is a specific numerical solution (integral) for a system of principal differential equations, which is of interest to us. In certain cases it is difficult to obtain such a relation by direct analytical means and so it is usually determined experimentally, i.e., by studying the phenomenon in the model with compulsory fulfilment of the main theorems of similarity and its additional postulates referred to in Sec. 1.3.

Even the experimental assay involves two major operations:

1. **Formulation of experiment** should be done in such a way that all quantities contained in the similarity criteria can be measured.
2. **Processing of experimental data** done in the criteria-related form, with presentation of results in the form of graphs for non-dimensional parameters (similarity criteria). The approximate experimental formulae that more satisfactorily describe the obtained criteria graphs are chosen.

However, this approach, which is the means for obtaining approximate formulae from the experiment, trails some major disadvantages, in particular:

1) Limited possibility of applying the method in toto due to inadequate number of criteria (two maximum of three criteria) used for plotting criteria-curves and relationships;

2) Limitation of approximate relationships obtained imposed by the intervals of variation in the governing criteria or parameters investigated in the experiment;

3) Discreteness of these relationships, i.e., existence of different relationships under different intervals of variation in the governing parameters for one and the same set of criteria.

An approximate analytical method also exists for determining these criterial relationships. It involves certain preliminary investigations concerning the role of individual similarity criteria in the phenomenon under study with the objective of identifying less influential or those not disturbing the similarity when criteria are altered in a particular range of values, i.e., in the range of automodellability of phenomena relative to these criteria. Then such similarity criteria may be excluded from the criterial equations and the resultant simplified equations readily solved analytically.

In some cases certain similarity criteria can be combined, by multiplying or dividing them or by adding or subtracting from them dimensionless numerical quantities. These can also be multiplied or divided by dimensionless numerical quantities. In all such cases new similarity criteria are obtained which, as a rule, retain the entire range of properties of the old while concomitantly simplifying the analytical determination of criterial equations.

Similarity criteria can be determined by two methods:

1) Based on analysis of a system of differential equations and conditions of uniqueness;
2) Based on dimensional analysis of the governing independent parameters.

It may be possible not to consider one or several factors (parameters) of the phenomena in the second case, while in the first a better guaranteed representation exists.

In the second case the similarity criteria are determined according to the laws of the π-theorem or dimension method as detailed above.

Here it should be mentioned that, apart from major units in the SI system (length—L, mass—M, time—T, current—I etc.), any other quantities x_1, x_2, $\dots x_K$ can be selected if the dimensions of such quantities are independent functions of the major units M, L, T, I and others in the SI system, i.e., $[x_1] \neq [x_2]^{\alpha} \dots [x_K]^{\beta}$ for any values of $\alpha, \dots, \beta$ and vice versa, if unique inverse transformation is possible, i.e., the dimensions of M, L, T, I and others can be uniquely expressed in terms of $[x_1], [x_2], \dots, [x_K]$.

Let us consider the conditions under which the above is satisfied. For the sake of simplicity a phenomenon in mechanics is taken, as only three major quantities are required for describing it: mass—M, length—L, time—T.

Let the dimensions of x_1, x_2, x_3 be equal to

$$[x_1] = M^{\alpha_1} L^{\beta_1} T^{\gamma_1};$$
$$[x_2] = M^{\alpha_2} L^{\beta_2} T^{\gamma_2};$$
$$[x_3] = M^{\alpha_3} L^{\beta_3} T^{\gamma_3},$$

where α_i, β_i, γ_i are arbitrary numerical coefficients ($i = 1, 2, 3$).

These logarithms are taken:

$$\log[x_1] = \alpha_1 \log M + \beta_1 \log L + \gamma_1 \log T,$$
$$\log[x_2] = \alpha_2 \log M + \beta_2 \log L + \gamma_2 \log T,$$
$$\log[x_3] = \alpha_3 \log M + \beta_3 \log L + \gamma_3 \log T.$$

It is known from courses in algebra that this system of equations has a unique solution if the determinant of the coefficients contained in the equations is not equal to zero, i.e.,

$$\Delta = \begin{vmatrix} \alpha_1 & \beta_1 & \gamma_1 \\ \alpha_2 & \beta_2 & \gamma_2 \\ \alpha_3 & \beta_3 & \gamma_3 \end{vmatrix} \neq 0. \qquad \ldots (1.30)$$

Actually, if an equality $[x_1] = [x_2]^{\lambda}[x_3]^{\tau}$ exists, where λ, τ are arbitrary numerical values, then taking the logarithms we obtain

$$\log[x_1] = \lambda \log[x_2] + \tau \log[x_3],$$

which means that in the first row of determinant (1.30) we have linear combinations of the remaining two quantities:

$$\alpha_1 = \lambda\alpha_2 + \tau\alpha_3; \ \beta_1 = \lambda\beta_2 + \tau\beta_3; \ \gamma_1 = \lambda\gamma_2 + \tau\gamma_3,$$

which leads to a condition contradicting (1.30), since the determinant becomes zero.

Thus condition (1.30) is the condition for possible selection of x_1, x_2, x_3 as key quantities. This rule obviously is characteristic even for a large number of principal units subject to the condition that the number of key quantities 'K' does not exceed the number of principal units in the SI system ($K = 6$ usually in problems involving mechanical, thermal, electrical and optical phenomena). Even in this case determinant (1.30) is of Kth order.

It has also been noted that the number of principal similarity criteria can be reduced by combining them through addition (subtraction), multiplication (division) etc., thereby obtaining more complex criteria, which in certain cases could simplify the formulation of model experiments.

First of all, they are readily and more reliably obtained as the governing parameters of the phenomenon are readily and reliably established, and the very process of finding similarity criteria amounts only to analysis of the system of differential equations. (It may be noted that the need for using similarity criteria in this instance, is associated with difficulties in mathematically solving for the system of equations and obtaining the integral solution.) Two methods for obtaining similarity criteria are differentiated: the method of similar transformations and the method of integral analogues.

The *first method* involves the following:

—System of principal differential equations and conditions of uniqueness are written.

—All variables, including the parameters, are replaced by similar proportional quantities, multiplying them by a corresponding similarity scale;

—Scale combinations of all the terms in equations thus obtained in compliance with the homogeneity of physical equations, are equated to each other and a system of specific equalities thereby obtained;

—Specific non-dimensional scale combinations are obtained by dividing one part of these equalities by the other. Such combinations are indicators of similarity amongst the unknown similitude criteria;

—Unknown similitude criteria would prove to be of the same type as the similarity indicators, if the quantities themselves are substituted by the scales of similarity.

The *second method* is simpler and hence more widely used. It is structured according to the rule of substitution whereby in the case of similar phenomena derivatives of any order of the characteristic quantities are replaced by ratios of corresponding quantities, i.e., the ratio of their so-called integral analogues. The method involves the following steps:

—Systems of differential equations and single-value equations are written;

—Simultaneously all the components contained in the vectors along the co-ordinate axes are substituted by their absolute values, the co-ordinates by the typical linear dimension of the system and the integral by expression under the integral;

—All the signs of equality of addition and subtraction are replaced by the sign of similarity or proportionality;

—As a result, certain combinations are obtained, consisting of the quantities (and not scales) themselves, which on equating to each other give unknown non-dimensional combinations. These are the similarity criteria sought.

After obtaining the similarity criteria by one method or the other, the physical phenomenon under study can be modelled. The following essential and sufficient conditions must be met if the phenomenon in the model is to duplicate that in nature:

1) model should be geometrically similar to the actuality (natural);
2) phenomena in the model and the actual should belong to the same group or class of phenomena, i.e., should be described by similar differential equations;
3) boundary conditions (conditions of uniqueness) in the model should be so realised that non-dimensional boundary conditions for the model and the actuality coincide entirely;
4) similarly termed non-dimensional parameters contained in the differential equations describing the model and the actuality should be correspondingly equal.

These requirements are related in toto to physical modelling. Let us now discuss in detail the fundamental concepts of physical modelling, which in essence lead us to solve the problem through experimental means: formulation and conduction of model experiment, processing of experimental data and extrapolation of data to the natural phenomenon under study. Here the model experiment should be clearly distinguished from a purely physical one. The difference primarily lies in the fact that in the model experiment the model serves as a substitute for the actual object under study. In a physical experiment the real object of investigation is subjected to various effects to determine its various physicomechanical characteristics.

Any physical experiment should be preceded by an analysis of equations of the physical phenomenon under study or analysis by the method of dimensionality of parameters and generalised co-ordinates featuring this phenomenon, so as to establish non-dimensional combinations from them which form the similarity criteria.

Further, the relationship between generalised co-ordinates and parameters in the criterial form is written in a general form, i.e., in the form of dependence of non-governing criteria (criteria which include generalised co-ordinates or quantities to be sought) on governing criteria (criteria formulated from parameters characterising the phenomenon or quantities satisfying conditions of uniqueness).

Later, based on the aforementioned conditions of modelling, a model is chosen and the essential requirements including the significant ranges for practical purposes or intervals of variation of governing parameters, and the essential and adequate accuracy of assigned and obtained quantities are worked out. Based on this, the procedure for conducting the experiment is developed and the measuring instrumentation selected. It should always be remembered that all parameters contained in all similarity criteria, both input (assigned) quantities and quantities to be sought, should be measured with a specified accuracy.

Then the experiments are conducted. The experiments are repeated twice-thrice under the same set of conditions to eliminate random errors in measurement and to obtain reliable values. After completion of experimentation, the most important part is processing the experimental data mathematically and obtaining empirical experimental relationships.

Thus, **physical modelling may be conditionally divided into four stages**:

1) analysis of phenomenon under study, selection of key parameters and generalised co-ordinates, determination of necessary and sufficient criteria of similarity;
2) evolving requirements for the model, scheme of model experiment and measuring apparatus;
3) preparation and conduction of model experiment;
4) processing of experimental data, its analysis and generalisation, and extrapolation of obtained results to the phenomenon being studied.

Let us discuss in detail the final or concluding stage in the model experiment.

To mathematically process the experimental data the form of equations interrelating the similarity criteria determined earlier by the methods of similarity criteria and dimensionality, should be concretely specified.

The step relationship, which is very simple and flexible, is taken as the specific form of relationship in the first approximation. This relationship is subsequently concretised while processing the experimental data.

The procedure for *mathematical processing of experimental data* is given below:

1. Let the governing parameters of two similarity criteria Π_1 and Π_2 and one unknown similarity criterion Π_x be obtained as a result of analysing the system of equations or by the dimension method. Π_x is related to Π_1 and Π_2 functionally as

$$\Pi_x = f(\Pi_1, \Pi_2).$$

In the first approximation let us assume f to be a stepped function, i.e.,

$$\Pi_x = C\Pi_1^n \Pi_2^m, \qquad \ldots (1.31)$$

where C is the coefficient of proportionality.

2. The experimental data obtained is presented in a criterial form, i.e., the numerical values of corresponding criteria Π_x, Π_1 and Π_2 are calculated on this basis.

3. Curves showing the relationship of Π_x and Π_1 at various values of Π_2 (co-ordinate axes $\log \Pi_x$, $\log \Pi_1$) or the relationship of Π_x and Π_2 at various values of Π_1 (co-ordinate axes $\log \Pi_x$, $\log \Pi_2$) are plotted, based on the experimental values of the criteria obtained, on a logarithmic scale.

4. If a stepped relation actually exists between Π_x and Π_1, the graph showing the relationship in the logarithmic scale will resemble an inclined straight line (Fig. 2, a):

$$\log \Pi_x = \log C_1 + n \log \Pi_1;$$
$$C_1 = C = \Pi_2^m.$$

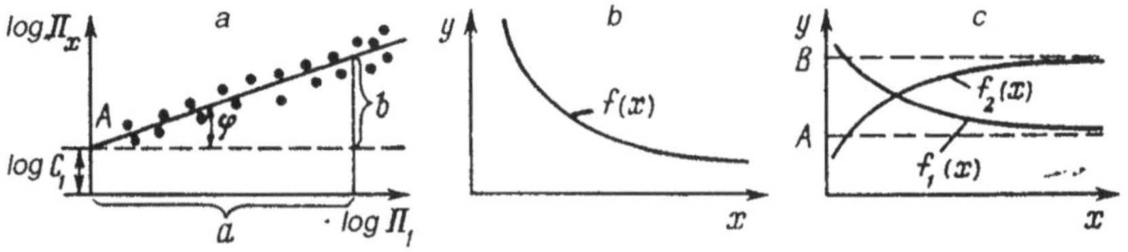

Fig. 2. Graphs showing relationship $\Pi_x = f(\Pi)_1$ in logarithmic scale (on both axes) in order to approximate it with (a) stepped function, (b) split linear function and (c) exponential function.

$$y = f(x) = e^{-\alpha x^n}; y = f_2(x) = B(1 - e^{-\alpha x^n})$$

where $y = \Pi_x, x = \Pi_1$.

The value of $\log C_1$ is taken directly from the graph on the y-axis at A, the point of intersection of the inclined straight line with this axis. The value of power index n is determined as the tangent of angle of inclination of the straight line with the x-axis, i.e.,

$$n = \tan\varphi = b/a.$$

The values of C_1 can also be computed based on the experimental data Π_x and Π_1, as the average value of an entire set of measurements Π_x and Π_1 at $\Pi_2 =$ const according to the formula

$$C_1 = C\Pi_2^m = \frac{1}{N}\sum_{p=1}^{p=N}\frac{\Pi_{x_p}}{\Pi_{1_p}},$$

where N is the number of times the values of Π_x and Π_1 are measured.

It is to be noted that the quantities C_1 and n may be computed according to the aforesaid formulae only when all the experimental points cluster along a straight line, which usually does not occur in practice. A certain scatter of experimental points is always observed and in such a case it is difficult to draw a straight line.

Hence it is better to determine the parameters of relationship (1.31) by the least squares method.

5. Subsequently a series of such graphs are plotted in a similar manner for various constant values of the criterion Π_2. The various values of C_1 obtained according to the method above for corresponding values of

$$C_1 = C\Pi_2^m$$

are used for determining the constant C and power index m by a similar method (if the graphs $\Pi_x = f(\Pi_1)$ are parallel).

Here it may so happen that even the power index n would vary with changing Π_2 (non-parallelism of graphs), i.e., $n = f(\Pi_2)$. This means that the accepted form of step-function (1.31) does not suit and, in this case, actually a more complex relationship exists between the criteria Π_x, Π_1 and Π_2.

6. If the experimental data processed according to the method cited under paragraph 4 do not lie on a straight line, this signifies the absence of a stepped relationship of the type (1.31) between the similarity criteria Π_x, Π_1 and Π_2. In such a case it is advisable to plot the graphs in the usual co-ordinate axes (Π_x, Π_1) or (Π_x, Π_2) and as per the obtained averaged curves to choose a matching empirical formula. In selecting the empirical formula, the following guidelines are to be followed:

a) For the inversely proportionate relationships (Fig. 2, b), $y = f(x)$, where $y = \Pi_x$, $x = \Pi_1$ (or Π_2), a piece-wise linear function is chosen, in particular:

$$\Pi_x = (A\Pi_1 + B)/(C\Pi_1 + D). \qquad \ldots (1.32)$$

b) In the case of a sharply changing curve (decreasing or increasing), asymptotically approaching a certain value (upper or lower), as shown in Fig. 2, c, the following exponential relationships are to be tried:

$$\text{for} \quad y_1 = f_1(x), \quad \Pi_x = A + e^{-\alpha \Pi_1^n}, \qquad \ldots (1.33)$$

$$\text{for} \quad y_2 = f_2(x), \quad \Pi_x = B(1 - e^{-\alpha \Pi_1^n}), \qquad \ldots (1.34)$$

where A and B are the corresponding asymptotic values (f_1 and f_2 in Fig. 2, c) of the unknown criterion $\Pi_x = y$; $x = \Pi_1$ is the governing criterion of similarity;

c) Should the curve oscillate within a certain period T, it is better to use trigonometric functions of the type

$$\Pi_x = A \sin\left(\alpha + \frac{2\Pi}{T}\Pi_1\right) \qquad \ldots (1.35)$$

or the series

$$\Pi_x = A_0 + A_1 \cos\frac{2\Pi}{T}\Pi_1 + A_2 \cos\frac{4\Pi}{T}\Pi_1 + \ldots$$
$$B_1 \sin\frac{2\Pi}{T}\Pi_1 + B_2 \sin\frac{2\Pi}{T}\Pi_1 + \ldots ;$$

d) If the shape of the curve does not allow selection of a particular formula, the formula of stepped polynomial of the following type should be used:

$$\Pi_x = A + B\Pi_1 + C\Pi_1^2 + \ldots + M\Pi_1^n.$$

In such a case the selection begins with a polynomial of 2nd degree, the relationship verified and, should it not prove suitable, the degree of the polynomial raised sequentially.

7. After selecting the type of empirical relationship, the numerical values of its symbolic parameters A, B, C, D and the power indices α, n, m are determined by the least squares method. Calculations are to be done in the following sequence [5]:

a) If the chosen relationship is non-linear relative to the criteria, it is initially brought into a linear form by a corresponding transformation or expansion into a series, in particular:
for a piece-wise linear relationship (1.32)

$$A\Pi_1 + B = (C\Pi_1 + D)\Pi_x;$$

for exponential relationships (1.33) and (1.34)

$$\ln[\ln(A/\Pi_x)] = m + n \ln \Pi_1;$$
$$\ln[\ln(\Pi_x/B) - 1] = m + n \ln \Pi_1,$$

where $m = \ln \alpha$;

for a trigonometric relationship (1.35)

$$\Pi_x = D\cos(2\pi\Pi_1/T) + C\sin(2\pi\Pi_1/T),$$
$$D = A\cos\alpha, C = A\sin\alpha.$$

b) Substituting in the thus-obtained linearised relationships, all the tabulated experimental values of criteria Π_x and Π_1 of S experiments, we obtain a series of linear conditional equations of the following type for determining the symbolic parameters A, B, C, $D\ldots$

$$f(\Pi_{1_i}, A, B, C, D\ldots) = \Pi_{x_i}, \qquad \ldots (1.36)$$

where $i = 1, 2, 3, \ldots, S$ (S is the number of experiments). They can be represented as the following relationships for finding A, B, C, D, ...

$$a_iA + b_iB + c_iC + dD\ldots = l_i,$$

where a_i, b_i, c_i, $d_i\ldots l_i$ are the coefficients of the conditional equations obtained as a result of substituting the tabulated values of Π_{1_i} and Π_{x_i} in (1.36); A, B, C, D are the unknown symbolic parameters of empirical formulae to be found.

c) To determine the unknown symbolic parameters, normal equations of the following type are written

$$A\Sigma a_i^2 + B\Sigma a_ib_i + C\Sigma a_ic_i + \ldots = \Sigma a_il_i,$$
$$A\Sigma a_ib_i + B\Sigma b_i^2 + C\Sigma b_ic_i + \ldots = \Sigma b_il_i,$$
$$A\Sigma a_ic_i + B\Sigma b_ic_i + C\Sigma c_i^2 + \ldots = \Sigma c_il_i,$$

where Σa_i^2, Σa_ib_i, Σa_il_i etc. are the sum of corresponding quantities a_i^2, a_ib_i, a_il_i etc. for all the S experiments, i.e., summation is done from $i = 1$ to $i = S$.

From this system of equations the unknown values of symbolic parameters A, B, C, D, α, n, m are determined.

8. Validity of computation of the coefficients of conditional linear equations is best verified before calculating the coefficients themselves, in the following manner:

a) The sum of coefficients of the conditional equations is determined, i.e.,

$$S_i = a_i + b_i + c_i + \ldots + l_i; \qquad \ldots (1.37)$$

b) the following equalities are verified:

$$\Sigma a_i^2 + \Sigma a_ib_i + \Sigma a_ic_i + \ldots + \Sigma a_il_i = \Sigma a_iS_i,$$
$$\ldots\ldots\ldots\ldots\ldots\ldots\ldots\ldots$$
$$\Sigma a_ic_i + \Sigma b_ic_i + \Sigma c_i^2 + \ldots + \Sigma c_il_i = \Sigma c_iS_i \qquad \ldots (1.38)$$

If (1.37) is satisfied, then the computation of parameters of the system (1.38) is taken up.

9. After obtaining the symbolic parameters and the empirical formulae, it is necessary to verify their dependability by doing the following calculations:

a) Substituting the tabulated values of governing criteria Π_1 in the formulae obtained; the computed values of unknown criteria are obtained Π_x.
b) From this data, the mean square error per unit weight is calculated

$$\delta = \sqrt{\Sigma(\Pi_x - \Pi_{x_p})^2/(s-k)},$$

or average absolute variance

$$\sigma = \Sigma(\Pi_x - \Pi_{x_p})/S,$$

where S is the number of tabulated values Π_x;
k is the number of parameters.

If the values of δ and σ are found to be within the absolute value of permissible errors of Π_x, the formula is said to be satisfactory, i.e.,

$$\Delta\Pi_x = \Pi_x - \Pi_{x_p} \leqslant \sigma, \quad \delta\Pi_x = \Delta\Pi_x/\Pi_x \leqslant \delta.$$

2

Application of Theories of Similarity and Dimension to Investigations into Blast Effects in Various Media

2.1 Hydrodynamic Similarity Criteria in Blasting in Water and Air Media

2.1.1 Similarity Criteria in Hydrodynamics

To derive similarity criteria in hydrodynamics let us consider a general case of liquid flow using the Navier-Stokes equation, which when presented on co-ordinate axes is of the following form:

$$\frac{\partial v_x}{\partial t} + v_x \frac{\partial v_x}{\partial x} + v_y \frac{\partial v_x}{\partial y} + v_z \frac{\partial v_x}{\partial z} =$$

$$X - \frac{1}{\rho}\frac{\partial p}{\partial x} + \frac{\nu}{3}\frac{\partial}{\partial x}(\text{div } \vec{v}) + \nu \Delta v_x;$$

. . . .

$$\frac{\partial v_z}{\partial t} + v_x \frac{\partial v_z}{\partial x} + v_y \frac{\partial v_z}{\partial y} + v_z \frac{\partial v_z}{\partial z} =$$

$$Z - \frac{1}{\rho}\frac{\partial p}{\partial z} + \frac{\nu}{3}\frac{\partial}{\partial z}(\text{div } \vec{v}) + \nu \Delta v_z, \qquad \ldots (2.1)$$

where $\text{div } \vec{v} = \frac{\partial v_x}{\partial x} + \frac{\partial v_y}{\partial y} + \frac{\partial v_z}{\partial z};$

$$\Delta v_i = \frac{\partial^2 v_i}{\partial x^2} + \frac{\partial^2 v_i}{\partial y^2} + \frac{\partial^2 v_i}{\partial z^2};$$

$$i = x, y, z; \; X = g_x; \; Y = g_y; \; Z = g_z.$$

Using the above equations, we shall find the similarity criteria by two methods: method of similar transformations and method of integral analogues.

In the *first method*, after choosing the scales of quantities to be contained in the equation, equation (2.1) is written for the model. By separating the scale complexes and equating the scale complexes thus obtained to each other, the similarity indicators are established. This procedure is adopted for the first equation of the system (2.1), and for the remaining two equations of the system, analogous similarity indicators are obtained.

Let us introduce the following scales of similarity:

$$l_N/l_M = l_c;\ t_N/t_M = t_c;\ v_N/v_M = v_c;$$
$$p_N/p_M = p_c;\ \rho_N/\rho_M = \rho_c;\ \nu_N/\nu_M = \nu_c;\ g_N/g_M = g_c. \quad \ldots (2.2)$$

The first equation of system (2.1) is transformed in accordance with (2.2).

$$\frac{v_c}{t_c}\left(\frac{\partial v_x}{\partial t}\right)_M + \frac{v_c^2}{l_c}\left[\left(v_x\frac{\partial v_x}{\partial x}\right)_M + \ldots + \left(v_z\frac{\partial v_x}{\partial z}\right)_M\right] =$$
$$g_c X_M - \frac{p_c}{\rho_c l_c}\left(\frac{1}{\rho}\frac{\partial p}{\partial x}\right)_M + \frac{\nu_c v_c}{l_c^2}\left[\frac{\nu}{3}\frac{\partial}{\partial x}(\text{div }\vec{v}) + \nu\Delta v_x\right]_M. \quad \ldots (2.3)$$

Equating the scale complexes of all the terms in (2.3),

$$v_c/t_c = v_c^2/l_c = g_c = p_c/(\rho_c l_c) = \nu_c v_c/l_c^2. \quad \ldots (2.4)$$

Dividing all the terms in (2.4) by one amongst them, for example, by v_c^2/l_c, the similarity indicators equal to unity are obtained:

$$l_c/(v_c t_c) = 1; \qquad g_c l_c/v_c^2 = 1;$$
$$p_c/(\rho_c v_c^2) = 1; \qquad \nu_c/(v_c l_c) = 1,$$

or by manipulating them,

$$v_c t_c/l_c = 1; \qquad v_c^2/(g_c l_c) = 1;$$
$$p_c/(\rho_c v_c^2) = 1; \qquad v_c l_c/\nu_c = 1. \quad \ldots (2.5)$$

Considering (2.2) and (2.5), the governing criteria of similarity would be:

$$v_N t_N/l_N = v_M t_M/l_M;$$

$vt/l =$ idem, homochronometry number (H) : (Strouhal Number);

$$v_N^2/(g_N l_N) = v_M^2/(g_M l_M);$$

$v^2/(gl) =$ idem, Froude number (Fr);

$$p_N/(\rho_N v_N^2) = p_M/(\rho_M v_M^2);$$

$p/(\rho v^2) =$ idem, Euler number (E);

$$v_N l_N/\nu_N = v_M l_M/\nu_M; \quad \ldots (2.6)$$

$vl/\nu =$ idem, Reynolds number (Re).

In the *second case* all the terms in the equation are substituted by their integral analogues and the conditions for similarity complexes, consisting of these integral analogues, are expressed, i.e.,

$$v/t \sim v^2/l \sim g \sim p(\rho l) \sim \nu v/l^2.$$

Equating them to each other and dividing by one amongst them, for example by v^2/l, the same similarity criteria are written in the final form (2.6).

From the similarity criteria thus obtained it transpires that flows of compressible viscous liquid would generally be similar if at identical points in the streams and at corresponding time periods the Strouhal, Euler, Froude and Reynolds numbers are equal.

It is well known that the patterns of flow of liquid and gas largely depend on the transport velocity of particles. In fluids, and particularly in gases, the velocity of sound exerts considerable influence. If the velocity of particles exceeds the sound velocity, then sharp densification and other peculiarities arise. Hence in gas dynamics one more similarity criterion—the Mach number—is widely used, which is a simplex of the type

$$v/c = M,$$

where c is the sound velocity.

Let us now discuss the significance of similarity criteria so obtained and the possibility of their simultaneous application.

Let us take, for example, the Froude and Reynolds numbers as they have significant applicational value, and because they provide simultaneous similarity of two governing forces: volumetric forces (gravitational forces) and viscous forces. It follows from the equality of these criteria that:

$$v_N l_N/\nu_N = v_M l_M/\nu_M, \text{ i.e., } v_M = v_N \nu_M l_c/\nu_N; \qquad \dots (2.7)$$

$$v_N^2/(g l_N) = v_M^2/(g l_M), \text{ i.e., } v_M = v_N \sqrt{1/l_c}; \qquad \dots (2.8)$$

here $g_M = g_N = g$.

Equating the right side of (2.7) and (2.8), we have

$$\nu_M/\nu_N = l_c^{-3/2}.$$

Usually $l_c > 1$ and consequently $\dfrac{\nu_N}{\nu_M} > 1$. This means that for the simultaneous implementation of the *Fr* and *Re* criteria, the kinematic viscosity of the fluid in the model stream should be less than the kinematic viscosity of fluid in the full-scale stream. This cannot be realised in practice, which is why the similarity criteria of *Fr* and *Re* cannot be satisfied simultaneously.

An analogous situation is often encountered while investigating other physical phenomena. Hence in most modelling cases a question arises concerning the partial fulfilment of similarity criteria, i.e., about approximate or incomplete modelling. Because of this the problem of errors in modelling also arises, which can be termed the scale effect and should be considered while converting (recalculating) the data of model experiments to the full-scale. Consideration of these errors is a fairly difficult task and is executed by additional analysis of parameters of the equation and, in certain cases, conducting additional special experiments.

Sometimes dissimilar phenomena are purposely modelled when certain similarity criteria Π_1 and Π_2 are not the same for the model and the full scale but the type of relationship between the unknown non-dimensional quantities and the criteria is known from certain other premises. Then the constancy of only such criteria on which dependence is not known is maintained while modelling.

Let us illustrate this type of modelling through vortex flow of liquid in a vertical pipe caused by rotation of an impeller near the water surface. This is illustrated schematically in Fig. 3.

By rotating an impeller with blades of diameter d, the fluid (water with soil particles) is set into motion in a cylindrical pipe of diameter D and length H. Consequently, vortex flow is generated with the vertical axis coinciding with the axis of impellers and pipes. This vortex flow of fluid, as is well known, leads to the formation of some degree of rarefaction inside the vortex, which increases as the source of vortex (impeller) is approached and a marked increase in the flow of fluid along the pipe axis towards the source occurs. This flow of fluid entrains soil particles from the bottom of the water tank and transports them to the impellers. While this is occurring, the pulp forms in the pipe due to the rotational-translational motion of the fluid.

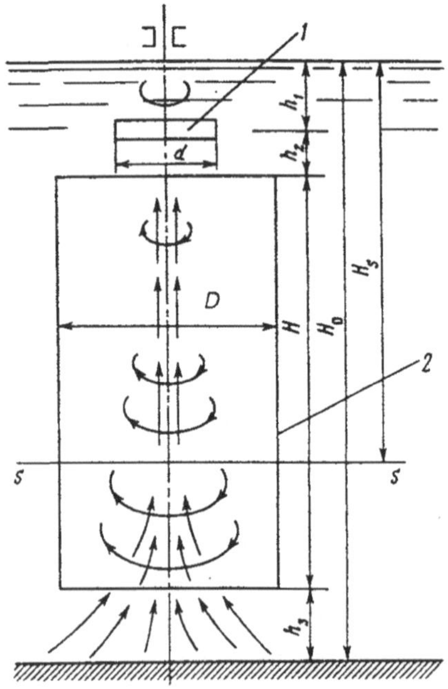

Fig. 3. Schematic diagram of vortex formation and ascending flow of liquid (pulp) in a pipe (2) when an impeller is rotated (1).

The pressure of rarefaction in a given section of the pipe (for a non-compressible and non-viscous fluid) [12] is given by:

$$p_s = p_{0s} - \rho\omega^2 D^2/4, \qquad \ldots (2.9)$$

where $p_{0s} = p_{atm} + \gamma H_s = p_{atm} + \rho g H_s$;
p_{0s} is the hydrostatic pressure in the given section of pipe s; ρ is the pulp density; ω is the angular velocity of rotation of pulp.

From (2.9) it is obvious that a reduction in p_{0s} occurs closer to the source while the second term on the right side increases rapidly due to increase in ω as the flow approaches the impeller. The vertically ascending flow of fluid along the axis of the vortex (pipe) is understood physically from the foregoing. This equation indicates the key parameters that govern the phenomenon under study, in particular:

ρ — density of pulp, $[\rho] = ML^{-3}$;
ω — angular velocity of fluid, $[\omega] = T^{-1}$;
D — diameter of pipe, $[D] = L$;
p_{atm} — atmospheric pressure, $[p_{atm}] = ML^{-1}T^2$;
g — acceleration due to gravity, $[g] = LT^{-2}$;
H_s —variable depth of loaded section, $[H_s] = L$ characterised by the depth of water tank H_0, lagging of pipe end from the bottom h_3, depth up to which the impeller is loaded h_1 and spacing between the blade and the upper edge of pipe h_2, i.e., by all the linear dimensions of the installation.

Obviously, the diameter of blade d should be added to the linear parameters, as it determines the assigned angular velocity ω of fluid.

Finally, as we are considering a real fluid and in particular a pulp consisting of mixtures of sea-water with soil particles, its viscosity ν, $[\nu] = L^2T^{-1}$, should also be taken into account.

It is common knowledge that viscosity substantially influences the flow of fluid at a high velocity near the walls of a pipe and surfaces of bodies.

Thus the quantity of vertical flow of fluid pulp, Q, $[Q] = L^3T^{-1}$ under a settled flow condition would be determined by the following parameters,

$$Q = f(\varphi, \omega, d, p_0, g, D, H_0, H_s, h_1, H, h_2, h_3, \nu). \qquad \ldots (2.10)$$

This functional relationship based on the π-theorem of similarity and the dimension theory can be replaced by the relationship between similarity criteria or by non-dimensional complexes composed of parameters of the process under study.

Taking ρ, ω and D as the key parameters, the unknown similarity criteria can readily be obtained using the dimension method. We have:

$$\Pi_1 = Q/(\omega D^3);\ \Pi_2 = p_0/(\rho\omega^2 D^2);\ \Pi_3 = g/(\omega^2 D);$$
$$\Pi_4 = \nu/(\omega D^2);\ \Pi_5 = H_0/D;\ \Pi_6 = h_i/D;$$
$$\Pi_7 = H/D;\ \Pi_8 = d/D;\ \Pi_9 = H_s/D. \qquad \ldots (2.11)$$

Let us reduce the number of similarity criteria (which is permitted by π-theorem), by combining the criteria Π_2, Π_3 and Π_5 in the following manner:

$$\Pi_2' = \Pi_2/(\Pi_3\Pi_5) = p_0/(\rho g H_0) = p_0/(\gamma H_0).$$

The similarity criteria $\Pi_6 - \Pi_9$ happen to be simplexes (ratios of similar quantities) of linear parameters and are automatically satisfied if geometric similarity is maintained.

Thus for modelling a given phenomenon, geometric similarity has to be satisfied and only two governing similarity criteria of the model and the full-scale should be the same,

$$\Pi_2' = p_0/(\gamma H_0) = p_0/(\gamma H_s) \text{ and } \Pi_4 = \nu/(\omega D^2). \qquad \ldots (2.12)$$

Functional relationship (2.10) is substituted by the following relationship between the governing criteria of similarity:

$$\Pi_1 = \varphi\left(\Pi_2', \Pi_4\right)$$

or

$$Q/(\omega D) = \varphi\left(p_0/(\gamma H_0);\ \nu/(\omega D^2)\right).$$

The governing criteria of similarity (2.11) can be obtained directly from the equation of Navier-Stokes for a steady-state flow of an incompressible viscous fluid in the vector form,

$$\text{rot } \vec{v}_x\omega = \vec{F} - \frac{1}{\rho} \text{ grad } p - \frac{1}{r} \text{ grad } \vec{v}^2 - \nu \text{ rot } \vec{\omega},$$

$$\vec{\omega}_x\vec{v} = g - \frac{1}{\rho} \text{ grad } p - \frac{1}{r} \text{ grad } \vec{v}^2 - \nu \text{ rot } \vec{\omega} \qquad \ldots (2.13)$$

Applying the method of integral analogues to equation (2.13), we obtain,

$$\omega v \sim g \sim p/(\rho l) \sim v^2/l \sim \nu\omega/l, \qquad \ldots (2.14)$$

where $l = r$ is the characteristic linear dimension.

Similarity criteria (2.13) can be readily obtained from equation (2.14) by dividing all the terms of similar equalities (2.14) by v^2/l, i.e.,

$$\omega l/v = gl/v^2 = p/(\rho v^2) = \nu\omega/v^2 = 1. \qquad \ldots (2.15)$$

From these equalities the governing criteria of similarity Π_2' and Π_4 are likewise easily obtained, by making the following simple transformations. Dividing the third term of equality (2.15) by the second term and dividing the fourth term by the square of the first, we obtain

$$\left[p/(\rho v^2)\right] / \left[gl/v^2\right] = p/\left[l\rho/g\right] = p/(\gamma l) = \Pi_2'';$$

$$\left[\nu\omega/v^2\right] / \left[v^2/(\omega^2 l^2)\right] = \nu/(\omega l^2) = \Pi_4.$$

The quantity of fluid flowing vertically along the pipe axis would be determined by the following criterial relationship

$$Q = \omega D^3 \varphi(p_0/(\gamma H_0), \nu/(\omega D^2)).$$

This indicates that modelling of the given phenomenon requires geometric similarity as well as equivalence of the governing criteria of similarity Π_2' and Π_4.

Let us discuss the conditions that arise out of the aforesaid governing criteria of similarity.

1) From the similarity criteria $\Pi_2' = p_0/(\gamma H_0)$ it follows that the following condition should be satisfied for both the model and the full-scale.

$$\frac{p_{0_N}}{\gamma_N H_{0_N}} = \frac{p_{0_M}}{\gamma_M H_{0_M}}$$

or $$p_{0_M} = p_{0_N} \frac{\gamma_M}{\gamma_N} \frac{H_{0_M}}{H_{0_N}} = p_{0_N} \frac{\gamma_M}{\gamma_N} \frac{1}{l_c} \qquad \ldots (2.16)$$

as from geometric similarity

$$H_N/H_M = D_N/D_M = h_{i_N}/h_{i_M} = d_N/d_M = l_N/l_M = l_c,$$

where l_c is the scale of geometric similarity, N and M are the indices for the full-scale and model respectively.

Condition (2.16) can be fulfilled in different ways.

1) If the same fluid is used in both the full-scale and the model, i.e.,

$$\gamma_N = \gamma_M,$$

then we have $p_{0_M} = p_{0_N}/l_c$.

This means that in this case the atmospheric pressure on water should be reduced proportional to the scale of geometric similarity. Thus if $l_c = 10^3$, then $p_{0_M} = 10^{-3} p_{0_N}$, i.e., a very large vacuum has to be created.

If the atmospheric pressure is not varied, i.e., $p_{0_N} = p_{0_M}$, then a heavy fluid has to be selected, i.e.,

$$\gamma_M = \gamma_N l_c \qquad \ldots (2.17)$$

or, in other words, the specific gravity of fluid in the model should exceed that of the full-scale by l_c times.

Equality (2.17) can also be written as

$$\gamma = \rho g, \text{ then } \rho_M g_M = \rho_N g_N l_c. \qquad \ldots (2.18)$$

If again the same liquid is selected (for both the model and the full-scale) i.e., $\rho_N = \rho_M$, then condition (2.18) is written as

$$g_M = g_N l_c,$$

which is the main condition for centrifugal modelling. In this case the model should be placed in a centrifuge and revolved in such a way that the centrifugal acceleration will exceed g_N by l_c times.

Finally, even a variant of the compromising solution is possible, which arises directly from condition (2.16),

$$\frac{p_{0_M}}{\gamma_M} = \frac{p_{0_N}}{\gamma_N}\frac{1}{l_c}, \qquad \ldots (2.19)$$

i.e., simultaneously creating a vacuum and using a heavier fluid, such that condition (2.19) is satisfied. This is usually adopted when l_c is large.

2) From the similarity criteria $\Pi_4 = \nu/(\omega D^2)$ the following condition for the model and the full-scale should be satisfied:

$$\nu_M/(\omega_M D_M^2) = \nu_N/(\omega_N D_N^2)$$

or

$$\nu_M = \nu_N \omega_M/(\omega_N l_c^2). \qquad \ldots (2.20)$$

This condition is very rigid and difficult to realise. Even if it is assumed that $\omega_M/\omega_N = 1/\omega_c = t_c = l_c$, where ω_c and t_c are the scales of similarity for ωt and time t respectively, it will result in one-time realisation of condition (2.11) (see criterion Π_3) and provide a less rigid condition

$$\nu_M = \nu_N/l_c.$$

From this condition, it follows that a heavy fluid has to be selected with $\gamma_M = \gamma_N l_c$ having very low viscosity, lower than that in the full-scale by

$$\nu_M = \nu_N/l_c.$$

This condition likewise is very difficult to realise. Consequently the exact modelling of any phenomenon under study is almost not possible. In such cases only approximate modelling can be adopted, for which certain preliminary investigations need to be undertaken on the dependence of the quantity of non-dimensional outflow of fluid $Q/(\omega D^3)$ on the governing criteria of similarity Π_2 and Π_4. The criterion Π_4 happens to be the most difficult to realise, which is why its influence on the outflow of fluid under study should be clarified first, i.e., the following relationship has to be studied,

$$\frac{Q}{\omega D^3} \approx f_1\left(\Pi_4 = \frac{\nu}{\omega D^2}\right). \qquad \ldots (2.21)$$

Relationship (2.21) can be investigated on the model by using several fluids (5–6 types of fluids, differing from each other by coefficients of kinematic viscosity). For example, nitrobenzol ($\nu = 0.0017$), acetone ($\nu = 0.00408$), methyl alcohol ($\nu = 0.00742$), water ($\nu = 0.0101$), linseed oil ($\nu = 0.038$), olive oil ($\nu = 0.094$) and glycerin ($\nu = 9.25$).

Tests have to be conducted at varying speed (at various ω), on either side of a suggested nominal value:

$$\omega_{nom_M} = \omega_{nom_N} l_c,$$

where ω_{nom_N} is the nominal value of ω in the full-scale; l_c is the suggested scale of geometric similarity.

While implementing the above test, if it appears that the values of $Q/\omega D^3$ for each constant ω vary from each other unsystematically when the type of fluid is changed, and the maximum relative range of variation does not exceed 15–20% for all ω (taken as parameters), then the influence of viscosity on the phenomenon under study may be neglected and only the first criterion Π_2 need be taken into account.

Most probably such a situation could arise only at a certain range of variation of ω (obviously at ω lesser than ω_{nom}). Consequently, approximate modelling can be done only in this range of variation of ω.

If there is a monotonous growth (fall) of relative outflow $Q/(\omega D^3)$ with change in viscosity of fluid, the influence of viscosity has to be considered and the fluid should be chosen according to condition (2.20).

As for the influence of similarity criterion Π_2, no doubts arise and so it does not call for any special investigation. The method of its realisation obviously would depend on results of investigations of the influence of similarity criterion Π_4 as considered above and technical possibilities in the laboratory.

It should be noted that the creation of a vacuum over the fluid would not only facilitate the realisation of criterion Π_2, but simultaneously reduce the value of kinematic viscosity of the selected fluid due to lowering of viscosity with a drop in pressure. In this connection, when criterion Π_4 exerts notable influence (influence of viscosity) and it is not possible to choose a corresponding fluid, the method of creating such a vacuum may be more useful and may be the only method possible for satisfying both the governing criteria of similarity in modelling the given phenomenon.

2.1.2 Similitude Criteria in the Theory of Blasting in Water and Air Media

In the theory of blasting in water and air [5], as is well known, equations of gas dynamics are used, in which forces of viscosity and mass (weight-related forces) are not considered. The Navier-Stokes equations (2.1) are simplified, mainly because only one term related to pressure remains in the right side (the second term). As a result, only two similarity criteria may be obtained from these equations, the Strouhal number and Euler number pertaining to kinematic and dynamic parameters of flow.

It follows from the first similarity criterion (Strouhal number) [see formula (2.6)] that:

$$v_N t_N / l_N = v_M t_M / l_M \rightarrow v_M = v_N t_c / l_c,$$

which indicates that velocity in similar streams would be equal if the condition

$$t_c / l_c = 1 \text{ or } t_c = l_c \qquad \ldots (2.22)$$

were satisfied, i.e., if the scales of linear dimensions and time are equal.

From the second criterion (Euler number) it follows that:

$$p_N / (\rho_N v_N^2) = p_M / (\rho_M v_M^2) \rightarrow p_M = p_N v_M^2 \rho_M / (v_N^2 \rho_N).$$

If condition (2.22) is satisfied and $v_M = v_N$, we have:

$$p_N = p_M\ \rho_M/\rho_N.$$

If the blast occurs in one and the same medium (in the model and the full-scale), i.e., $\rho_N = \rho_M$, then even the pressures are observed to be equal $p_N = p_M$.

Thus pressure and velocity of fluid (gas) particles during blasts in the same medium would be the same at analogous spatial points and at analogous time periods if the scales of time and linear dimensions are the same.

On the other hand, from the condition of dynamic compatibility that interrelates the parameters of undisturbed and disturbed fluid and from the general propositions of the detonation theory, it is known that initial values of pressure and velocity of particles at the shock front are determined by the explosive composition (*EC*) and properties of the surrounding medium.

It thus follows that in blasts of charges containing the same *EC* and in the same medium, pressure and velocity of fluid (gas) particles would be equal at analogous spatial points and at analogous time periods, if the scales of time and linear dimensions [condition (2.22)] are maintained the same.

A charge dimension, usually the radius of a spherical or cylindrical charge, is taken as a characteristic linear dimension in the theory of blasting. The *similitude law in blasts under water and in air* is formulated thusly: parameters of the medium (pressure and velocity of particles) during the flow caused by firing the same *EC* do not change if the scales of length and time by which these parameters are measured, vary by the same number of times as the charge dimensions.

Thus pressure in the shock wave, measured at a distance r from the spherical charge of radius R_{0_z}, should be equal to the pressure in the shock wave at a distance $r' = \lambda' r$ from a charge with a radius of $R'_{0_z} = \lambda' R_{0_z}$.

The shock-wave diagram (pressure-time) would be of the type shown in Fig. 4.

Here equal time periods t_1, t_2, t_3 and t_4 are taken, i.e., $t_2 = 2t_1, t_3 = 3t_1;\ t_4 = 4t_1$ and $\lambda = (\lambda' - 1)\Delta t$, where $\Delta t = t_1$.

It can be seen from Fig. 4 that the analogous spatial points in the theory of blasting are those which have equal ratios between distances from the charges and the corresponding charge radii (r/R_{0_z}) and analogous time periods are those, which have equal ratios between time periods and corresponding charge radii (t/R_{0_z}). In certain cases the non-dimensional ratio is used in place of t/R_{0_z},

$$\bar{r} = ta_0/R_{0_z},$$

where a_0 is the velocity of sound in fluid or air.

In place of charge radius, to express analogous spatial points, in certain cases the cube root of weight of spherical or concentrated charge $\sqrt[3]{G}$ or square root of weight of 1 m long extended (cylindrical) charge $\sqrt{q}$ is used,

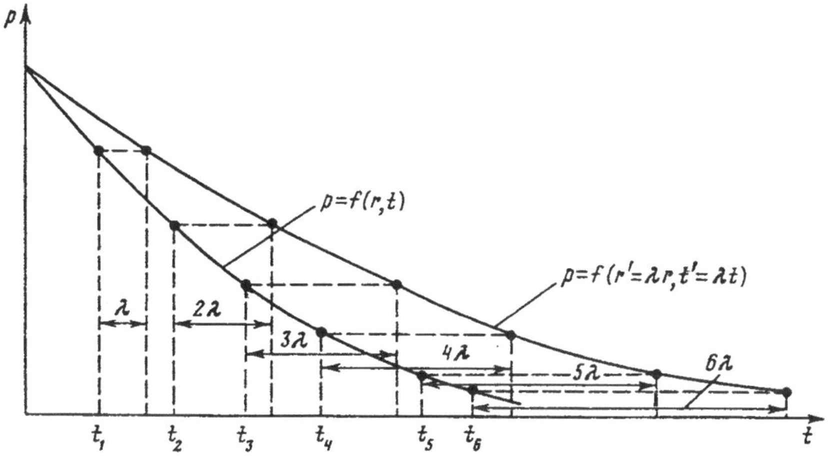

Fig. 4. Variation of pressure in shock wave over time at distances r and $r' = \lambda r$ in blasts of charges with radii R_{0_z} and $R'_{0_z} = \lambda R_{0_z}$ respectively.

since for one and the same *EC* there exists a direct proportionality between radius and charge weight, in particular:

for a spherical charge $R_{0_z} = C_1 \sqrt[3]{G}$; $C_1 = \sqrt[3]{3/4\pi\gamma_{EC}}$,

for an extended charge $R_{0_z} = C_2\sqrt{q}$; $C_2 = \sqrt{1/\pi\gamma_{EC}}$.

In this case, in place of relative distances $\bar{r} = r/R_{0_z}$ and relative times t/R_{0_z}, the so-called scaled distances $r/\sqrt[3]{G}$ or $r/\sqrt{q}$ and scaled time $t/\sqrt[3]{G}$ or $t/\sqrt{q}$ are introduced.

Formulation of the similitude law becomes somewhat complicated if the hydrodynamic fields of fluid generated in blasts of different *EC* are compared. The energy-based principle has to be relied upon as the basis for similarity. This principle is justified for specific distances from the charge at which the process of intensive dissipation of blast energy would have already been completed. But evidently at very close proximity to the charge, these [hydrodynamic] fields are rather dissimilar due to differences in boundary conditions.

Based on this principle, analogous spatial points and analogous time periods are determined from the relation:

$$r_1/\sqrt[3]{E_1} = r_2/\sqrt[3]{E_2};\ t_1/\sqrt[3]{E_1} = t_2/\sqrt[3]{E_2}, \qquad \ldots (2.23)$$

where E_1 and E_2 are the blast energy of two different *ECs*.

The blast energy of any industrial *EC* is proportional to the charge weight G and the amount of heat transformation Q due to the blast, i.e., $E = GQ$. Expressing the charge weight in terms of density ρ and volume V, the energy of spherical and concentrated charges may be written as

$$E = \rho_{EC} VQ = 4\pi R_{0_z}^3 \rho_{EC} Q/3. \quad \ldots (2.24)$$

The corresponding expression for the energy of unit length of extended and cylindrical charges would be

$$E_0 = \rho_{EC} V_0 Q = \pi R_{0_z}^2 \rho_{EC} Q. \quad \ldots (2.25)$$

Here the charge radii are determined by expressions (2.24) and (2.25) for spherical and cylindrical charges respectively, while the concentrated charges of arbitrary form are reduced to the spherical form by the expression

$$R_{0_z} = \sqrt[3]{3V/4\pi}.$$

where R_{0_z} would be expressed in m if weight of spherical charge is in kg, specific weight of cylindrical charge q in kg/m and volume of concentrated charge V in m^3.

Based on formulae (2.24) and (2.25), ratio (2.23) can be rewritten as: for spherical charges

$$r_1/\left(R_{0_{z_1}} \sqrt[3]{\rho_{EC_1} Q_1}\right) = r_2/\left(R_{0_{z_2}} \sqrt[3]{\rho_{EC_2} Q_2}\right);$$

$$t_1/\left(R_{0_{z_1}} \sqrt[3]{\rho_{EC_1} Q_1}\right) = t_2/\left(R_{0_{z_2}} \sqrt[3]{\rho_{EC_2} Q_2}\right); \quad \ldots (2.26)$$

for cylindrical charges

$$r_1/\left(R_{0_{z_1}} \sqrt{\rho_{EC_1} Q_1}\right) = r_2/\left(R_{0_{z_2}} \sqrt{\rho_{EC_2} Q_2}\right);$$

$$t_1/\left(R_{0_{z_1}} \sqrt{Q_1 \rho_{EC_1}}\right) = t_2/\left(R_{0_{z_2}} \sqrt{\rho_{EC_2} Q_2}\right) \quad \ldots (2.27)$$

By introducing the concept of equivalent charge radius based on blast energy, the above expressions may be written more, simply as

$$r_1/R_{0_{z_1}} = r_2/R_{0_{z_2}}^*; \; t_1/R_{0_{z_1}} = t_2/R_{0_{z_2}}^*,$$

where $R_{0_{z_2}}^*$ is the radius of charge containing the second EC, equivalent to the radius of charge with first EC, based on the blast energy.

The equivalent charge radius according to (2.26) and (2.27) is given by: for a spherical charge

$$R_{0_{z_2}}^* = R_{0_{z_2}} \sqrt[3]{\rho_{EC_2} Q_2/(\rho_{EC_1} Q_1)}; \quad \ldots (2.28)$$

for a cylindrical charge

$$R_{0_{z_2}}^* = R_{0_{z_2}} \sqrt{\rho_{EC_2} Q_2/(\rho_{EC_1} Q_1)}. \quad \ldots (2.29)$$

In certain cases the scaled distances and scaled time periods are determined based on the blast energy equivalent to the charge weight G^* or q^*.

Analogous to the aforementioned discussions and from (2.23) we have,

$$r_1/\sqrt[3]{G_1Q_1} = r_2/\sqrt[3]{G_2Q_2} \text{ or } r_1/\sqrt[3]{G_1} = r_2/\sqrt[3]{G_2^*},$$

$$r_1/\sqrt{Q_1q_1} = r_2/\sqrt{Q_2q_2} \text{ or } r_1/\sqrt{q_1} = r_2/\sqrt{q_2^*}.$$

Hence the weight of a second charge equivalent to the weight of the first charge based on blast energy, would correspondingly be equal to:

$$G_2^* = G_2Q_2/Q, \text{ or } q_2^* = q_2Q_2/Q_1.$$

In the theory of blasting, Trotyl* with a density of 1600 kg/m^3 and heat of blast transformation $Q = 4440$ kJ/kg, is taken as the standard *EC*. Thus we speak of a Trotyl-based charge weight, meaning thereby that weight of a Trotyl charge whose initiation gives the same effect as that obtained by firing an explosive charge under comparison. We similarly speak of a Trotyl charge weight based on a shock wave (of interest in our case) and also on seismic effects or other blast-induced effects.

The radius of spherical (and also the concentrated) standard Trotyl charge is calculated using an approximate formula

$$R_{0_z} = 0.053\sqrt[3]{G}, \qquad \ldots (2.30)$$

where G charge weight, kg.

The radius of a charge of any other *EC* equivalent in blast energy to a Trotyl charge, can be determined by formulae (2.28) and (2.29) in which $\rho_{EC_1} = 1600$ kg/m^3 and $Q_1 = 4440$ kJ/kg. It can also be determined by a simpler relationship (concentrated and spherical charges) by introducing a weight of charge equal to Trotyl into formula (2.30), i.e., according to the relationship:

$$R_{0_z}^* = 0.053\sqrt[3]{G_{EC}^*};\ G_{EC}^* = G_{EC}Q_{EC}/4440 \qquad \ldots (2.31)$$

Thus, after introducing the concept of radius of charge of equivalent blast energy, let us revert to the law of geometric similarity formulated earlier for a single *EC* and later extended to all types of *EC*, i.e., applying the so-called *extended law of geometric similarity*. This law can be formulated as follows: parameters of the medium (pressure, density and velocity of particles) set in motion by the blast do not change if the scales of length and time by which the parameters are measured, are scaled down to the same extent as the radius of charge with equivalent blast energy.

According to the law of extended geometric similarity, the formula for the pressure developed in a blast conducted in an unconfined fluid may be written in the form,

---

* Chemical formula: $C_6H_2(NO_2)_3CH_3$—Translator.

$$p = C_1 f\left(r/R^*_{0_z}, t/R^*_{0_z}\right)$$

or

$$p = C_2 f\left(r/R^*_{0_z};\ ta_0/R^*_{0_z}\right).$$

Obviously, for maximum pressure at the shock front the equalities given below are justified:

$$p_{max} = C_3 f\left(r/R^*_{0_z}\right);\ p_{max} = C_4 f\left(r/\sqrt[3]{G^*}\right) \quad \ldots (2.32)$$

In these formulae C_1, C_2, C_3, C_4 are certain dimensional constants.

The expression for the pulse of shock wave, a very important characteristic that determines the impact of a shock wave on any barrier and structure, is given below:

$$J = \int_0^T p\, dt = C_5 \int_0^T f\left(r/R^*_{0_z};\ a_0 t/R^*_{0_z}\right) dt$$

$$= C_5 R^*_{0_z}/a_0 \int_0^T f\left(r/R^*_{0_z};\ a_0 t/R^*_{0_z}\right) d\left(a_0 t/R^*_{0_z}\right),$$

where C_5 is a dimensional constant; a_0 the velocity of sound in a fluid.

In the above expression the non-dimensional parameters are under the integral. Consequently, at the same relative distance the magnitude of shock-wave pulses would be proportional to the charge radii.

Arising out of the first condition of dynamic correspondence, the velocity of fluid particles at the shock front may also be expressed analogous to the expressions derived for pressure:

$$p_{max} = \rho_0 a_0 v_{max},$$

where ρ_0, a_0 are density of undisturbed medium and sound velocity in it respectively; v_{max} the velocity of part of the medium at the shock front.

It should be noted that the similitude law developed above, often described as the extended law of geometric similarity, is in essence the law of energy-related similarity and is justified over almost the entire area, except that immediately surrounding the charge, since due to differences in boundary conditions for different *EC*, similarity is not observed in such fields.

A more generalised approach is possible for selecting the system of non-dimensional parameters in the near-field zone of blast. Such parameters characterise a powerful blast with very large parameters at the shock front (mainly p and θ), compared to those in undisturbed medium. This approach was developed on the basis of the dimension theory by L.I. Sedov for air shock waves.

2.2 The Problem of a Powerful Blast (L.I. Sedov's Solution)

Let a blast be conducted in a stationary gaseous medium at a certain point (in the centre of symmetry) at moment $t = 0$, i.e., energy E_0 is suddenly released. Let us neglect the weight and dimensions of the *EC* which are characteristic of an atomic explosion (weightless explosion) or electrical discharge (lightning).

Presuming that the key parameter for characterising a blast is the blast energy E_0, and that the properties of the medium are adequately determined by initial density ρ_0, pressure p_0 and coefficient of adiabatic expansion k, L.I. Sedov focussed attention on the following governing parameters for assessing the adiabatic movement of gas: $E_0, \rho_0, p_0, k, r, t$, where r is distance from the blast centre; t the time elapsed from the moment of blast.

The above quantities have the following dimensions:

$$[E_0] = ML^{\nu-1}T^2, \qquad [\rho_0] = ML^{-3},$$
$$[p_0] = ML^{-1}T^2, \qquad [k] = 1, \qquad [r] = L, \qquad [t] = T.$$

Here $\nu = 1, 2, 3$ corresponding to motion with planar, cylindrical and spherical symmetry. In writing the dimension of energy of motion with cylindrical symmetry, the quantum of energy is scaled to unit length, and for motion with planar symmetry, reduced to unit surface.

Of the five dimensional quantities E_0, ρ_0, p_0, r and t, three, namely ρ_0, r, t, have independent dimensions. According to the π-theorem only $(5-3) = 2$ non-dimensional combinations may be derived from the principal quantities as well as the remaining E_0 and p_0. Such combinations would be

$$\Pi_1 = E_0 t^2/(\rho_0 r^{\nu+2}), \ \Pi_2 = p_0 t^2/(\rho_0 r^2).$$

The third similarity criterion, which is equal to the non-dimensional coefficient of adiabatic expansion k, i.e., $\Pi_3 = k$, should be added to the above in order to obtain total similarity.

These similarity criteria determine completely the process of propagation of a shock front in the blasting of a point charge.

In the criteria Π_1 and Π_2 two variable quantities r and t occur. Generally, the π-theorem permits ambiguous selection of governing criteria of similitude. For example, it may be required that there should be no dependence on r in the second criterion, i.e., it should only be dependent on the variable t.

This is easy to obtain if Π_2 is represented as $\Pi_2' = t/\rho_0^x E_0^y p_0^z$, since then we get $\Pi_2' = t p_0^{1/2+1/\nu}/(E^{1/\nu}\rho^{1/2})$.

The second criterion of L.I. Sedov is obtained in this manner.

Thus the parameters of the medium at a propagating shock front in a powerful blast were determined according to L.I. Sedov by three similarity criteria

$$\Pi_1 = k, \Pi_2 = \lambda = E_0 t^2 / \left(\rho_0 r^{2+\nu}\right),$$
$$\Pi_3 = \tau = t p_0^{1/2+1/\nu} / \left(E_0^{1/\nu} \rho_0^{1/2}\right). \qquad \ldots (2.33)$$

As a powerful blast is being studied, in the near-field zone, the pressure at the front is considerably more than the initial (by 3–4 times or more). Consequently, the initial pressure of the gas and, correspondingly, the variable τ, may be neglected in this case. Then motion would be determined only by a single non-dimensional parameter λ. From the structure of λ it follows that the process can be considered an automodel (automatically prone to modelling), because it depends only on simple variables: distances r and time t at known E_0 and p_0.

Let us consider the hydrodynamic parameters at the shock front caused by a powerful blast: pressure p_f, density ρ_f and particle velocity v_f. It is known that the values of these parameters are discontinuous at the shock front, sharply rising from p_0, ρ_0, v_0 to p_f, ρ_f, and v_f. Their interrelationship is determined by the so-called conditions of dynamic correspondence, which are nothing more than a mathematical form of description of the general law of conservation of quantum of motion, continuity and energy. These conditions were obtained by N.E. Kochin in the following form:

$$[\rho\theta] = 0,$$
$$\rho\theta[\vec{v}] = [p]\vec{n},$$
$$\rho\theta\left[v^2/2 + u/A\right] = [pv_n], \qquad \ldots (2.34)$$

where $[p] = p_- - p_+ = p_f - p_0$; $\theta = N - v$ the velocity of displacement of shock front; N the velocity of propagation of shock front; v the velocity of particles of the medium; A the thermal equivalent of work done; u the internal energy.

These conditions of dynamic correspondence (2.34) for an ideal gas may be represented in the following form after a series of transformations:

$$\frac{\rho_f}{\rho_0} = \frac{\dfrac{k+1}{k-1}\dfrac{p_f}{p_0} + 1}{\dfrac{k+1}{k-1} + \dfrac{p_f}{p_0}},$$

$$N = a_0\sqrt{\frac{k+1}{2k}\frac{p_f}{p_0} + \frac{k-1}{2k}},$$

$$v_f = \frac{a_0\left(\dfrac{p_f}{p_0} - 1\right)}{k\sqrt{\dfrac{k+1}{2k}\dfrac{p_f}{p_0} + \dfrac{k-1}{2k}}}, \qquad \ldots (2.35)$$

where a_0—velocity of sound in undisturbed medium.

Assuming $p_f/p_0 \gg 1$ from (2.35) we get

$$\rho_f = \frac{k+1}{k-1}\rho_0, \qquad p_f = \frac{2}{k+1}\rho_0 N^2,$$

$$v_f = N\frac{\frac{k+1}{k-1}\rho_0 - \rho_0}{\frac{k+1}{k-1}\rho_0} = \frac{2}{k+1}N. \qquad \ldots (2.36)$$

From the constancy of density at the wavefront [(refer to the first equality from 2.36)], it follows that even the parameter λ according to (2.33) should have a constant value $\lambda = \lambda^*$. The law of motion of shock wave in such a case is readily established directly from the criterion Π_2 according to (2.33), substituting r, t, ρ_0, in it by r_f, t_f, ρ_f, respectively, namely:

$$r_f = \left(\frac{E_0}{\rho_f \lambda^*}\right)^{\frac{1}{2+\nu}} t_f^{\frac{2}{2+\nu}}.$$

Differentiating this with respect to time,

$$\frac{dr_f}{dt} = N = \frac{2}{2+\nu}\left(\frac{E_0}{\rho_f \lambda^*}\right)^{\frac{1}{2+\nu}} t_f^{-\frac{\nu}{2+\nu}} = \frac{2r_f}{2+\nu t_f}. \qquad \ldots (2.37)$$

This relationship accords well with the data of trial atomic explosions.

Using (2.37), the equations of dynamic correspondence for spherical symmetry $\nu = 3$ may be written as:

$$\rho_f = \frac{k+1}{k-1}\rho_0;$$

$$p_f = \frac{8}{(2+\nu)^2}\frac{\rho_0}{k+1}\left(\frac{r_f}{t_f}\right)^2 = \left|\frac{8}{25}\frac{\rho_0}{k+1}\left(\frac{r_f}{t_f}\right)^2\right|_{\nu=3};$$

$$v_f = \frac{4}{(2+\nu)(k+1)}\left(\frac{r_f}{t_f}\right) = \left|\frac{4}{5(k+1)}\left(\frac{r_f}{t_f}\right)\right|_{\nu=3}.$$

The parameters at the front of a powerful shock wave, namely p_f, ρ_f, v_f can be entirely determined by using the above system of equations.

Let us now consider the parameters of an ideal gas at the rear of a shock front. It is common knowledge that they are determined by the system of differential equations comprising equations of motion (from the law of conservation of quantum of motion), equations of continuity (from the law of conservation of energy) and equations of the state of medium in which the blast occurs. This system for unidimensional adiabatic motion of ideal gas would be of the form:

$$\frac{\partial v}{\partial t} + v\frac{\partial v}{\partial r} + \frac{1}{\rho}\frac{\partial p}{\partial t} = 0,$$
$$\frac{\partial p}{\partial t} + \rho\frac{\partial v}{\partial r} + v\frac{\partial \rho}{\partial r} + (\nu - 1)\frac{\rho v}{r} = 0,$$
$$\frac{\partial}{\partial t}\left(\frac{p}{\rho^k}\right) + v\frac{\partial}{\partial r}\left(\frac{p}{\partial^k}\right) = 0. \qquad \ldots (2.38)$$

Let us introduce the dimensionless functions of velocity $V(\lambda)$, density $R(\lambda)$ and pressure $P(\lambda)$, designating them by the following relationships:

$$v = \frac{r}{t}V(\lambda); \qquad \rho = \rho_0 R(\lambda); \qquad p = \rho_0\frac{r^2}{t^2}P(\lambda) \qquad \ldots (2.39)$$

Paying attention to the following

$$\frac{\partial}{\partial t} = \frac{\partial}{\partial \lambda}\frac{\partial \lambda}{\partial t} = 2\frac{E_0}{\rho_0}\frac{t}{r^{2+\nu}}\frac{d}{d\lambda} = 2\frac{\lambda}{t}\frac{d}{d\lambda};$$
$$\frac{\partial}{\partial r} = \frac{\partial}{\partial \lambda}\frac{\partial \lambda}{\partial r} = -(2+\nu)\frac{E_0}{\rho_0}\frac{t^2}{r^{3+\nu}}\frac{d}{d\lambda} = -(2+\nu)\frac{\lambda d}{2d\lambda}.$$

after substituting the new variables V, p and R in equation (2.38), we obtain a specific system of equations for them:

$$\lambda\left[(m - V)V' + m\frac{P'}{R}\right] = V^2 - V + \frac{2P}{R},$$
$$\lambda\left[mv' + (mv - 2)\frac{R'}{R}\right] = \nu V,$$
$$\lambda(mv - 2)\left[\frac{P'}{P} - k\frac{R'}{R}\right] = 2(V - 1), \qquad \ldots (2.40)$$

where $m = 2 + \nu$.

It is possible to determine the functions V, R and P by integrating the above system of equations, using the conditions at the shock front as the initial data. Later, it is easy to evaluate the parameters at the rear of the wave front v, ρ and p based on relationships (2.39). However, (2.40) can be integrated only by numerical methods as its integration in the final form is not possible.

L.I. Sedov suggested a special solution that enables one to find the algebraic integral of system (2.40) based on the propositions of the dimension theory.

With this in view, variation in the total energy of gas in a certain volume Q, confined by a movable spherical (cylindrical, flat) surface Σ ($\lambda = \text{const}$), is computed by the expression

$$\frac{dE}{dt} = \frac{d}{dt} \int_Q \left(\frac{\rho v^2}{2} + \rho\varepsilon \right) dq,$$

where ε—internal energy of gas

$$\varepsilon = p/[(k-1)\rho]. \qquad \ldots (2.41)$$

Let r be the co-ordinate that characterises the volume Q or the surface Σ surrounding it. Expanding according to law (2.37), the surface $\lambda =$ const during time dt causes an increment in the volume $4\pi r^2 N dt$ in which gas having energy $4\pi r^2 N dt \left(\frac{\rho v^2}{2} + \rho\varepsilon \right)$ is confined. This energy should be equal to the energy $4\pi r^2 v dt \left(\frac{\rho v^2}{2} + \rho\varepsilon \right)$ being carried by the gas flowing through a sphere of radius r during time dt and to the work done by forces of pressure $4\pi r^2 p v dt$.

Thus we have:

$$4\pi r^2 N dt \left(\frac{\rho v^2}{2} + \rho\varepsilon \right) = 4\pi r^2 v dt \left(\frac{\rho v^2}{2} + \rho\varepsilon \right) + 4\pi r^2 p v dt$$

or, simplifying

$$N \left(\frac{\rho v^2}{2} + \rho\varepsilon \right) = v \left(\frac{\rho v^2}{2} + \rho\varepsilon + p \right).$$

After substituting the values of N from (2.37) and ε from (2.41), we get

$$\frac{2r}{(2+\nu)t} \left(\frac{\rho v^2}{2} + \frac{p}{k+1} \right) = v \left(\frac{\rho v^2}{2} + \frac{k}{k-1} p \right).$$

Replacing the quantities v, ρ and p in the above equation by dimensionless quantities V, R and P, and according to formulae (2.39), we finally get an integral for system (2.40):

$$RV^2 \left(V - \frac{2}{2+\nu} \right) + \frac{2}{k-1} P \left(kV - \frac{2}{2+\nu} \right) = 0.$$

Using this relationship, the final solutions for V, R and P can be obtained, which are given in the book of L.I. Sedov, *Theory of Similitude and Dimensionality in Mechanics* (Nedra, Moscow, 1972).

2.3 Specific Features of Similarity Criteria in the Construction of Buildings Subjected to Explosion Indirectly

The strength of any structure exposed indirectly to a blast should obviously be computed based on the effect of blast-induced shock wave or stress wave on the structure. It is determined by the propagation of elastoplastic waves arising in the structure as a result of refraction and reflection of incident shock or stress waves and their diffraction around it.

The processes of diffraction, reflection and refraction of a blast wave as well as the process of initiation and propagation of elastoplastic waves in the structure, depend largely on the type of structure itself, its configuration and material, supporting braces and other factors. Owing to the flexibility of certain elements of construction, additional diffractional fields are created, while in water the phenomenon of cavitation is observed, which further complicates the wave picture around the structure. Three types of elastoplastic waves are transferred in the elements of the structure: longitudinal, transverse and surface. Moreover, multiple reflection and refraction further complicate the wave picture. Consequently a study of the indirect, non-contacting effect of a blast on a structure in terms of accurate wave angle is almost impossible, even with the most modern mathematical methods.

Usually, while solving practical problems certain simplifying assumptions and allowances are introduced in investigations carried out by means of modelling.

All the varied and complex phenomena associated with the effect of non-contact blasting on a structure may be characterised by the following governing parameters, for example, for the effect of blast on a plate of finite dimensions:

$$P, R_{0_z}, \rho_p, c_p, \rho_k, c_k, \sigma_k, \delta_k, E, \nu, y, y', y'', r, t,$$

where ρ_k, c_k are density and velocity of sound in the structure; δ_k plate thickness; y, y', y'' linear deformation of plate and its derivatives; σ_k stress in the structure; P certain resultant load, proportional to the square of the modulus of geometric similarity (linear scale).

A functional relationship in the criterial form for deflection of the plate can be obtained using the π-theorem.

$$y/R_{0_z} = f\left(r/R_{0_z}, \delta_k/R_{0_z}, \rho_p c_p/(\rho_k c_k), \sigma_k/(\rho_p c_p^2), \nu P/(E\delta_k^2), c_p t/R_{0_z}\right).$$

The similarity criteria $y/R_{0_z}, r/R_{0_z}, \delta_k/P_{0_z}$ hold good when geometric similarity is satisfied.

The criterion $\sigma_k/(\rho_p c_p^2)$ specifies the property of the medium surrounding the structure and is automatically fulfilled if the blast is conducted in one and the same medium, for example, in water or homogeneous rock.

The criterion $\rho_p c_p/(\rho_k c_k)$ denotes the similarity of reflection and refraction of waves at similar points and is automatically fulfilled if the medium and material used for construction in the model and the full-scale are the same.

The criterion ν = const characterises the elastic properties of the material used for construction and, obviously, is also fulfilled when the same material is used in both structures.

The criterion $P/(E\delta_k^2)$ denotes the nature of resultant load, which is usually fulfilled automatically if the same materials are used in the model and the full-scale.

The criterion $c_p t/R_{0_z}$ establishes the similarity of time characteristics and requires that the time parameters be proportional to the radii of charges, periods of natural vibrations of structural elements, duration of action of forces and time-related nature of the effect (diagrams of pressure-time; stress-time etc.).

The similarity criteria specify the general conditions of similarity for solid deformable bodies under the dynamic effect of external forces, without concretising the nature of application of such forces.

However, the nature of application of external load, namely body (volumetric) force, concentrated force etc., exerts considerable influence on the modelling of elements of deformation in structures and the internal stresses in them.

Generally speaking, the external load acting on any structure is distinctive—concentrated; distributed over length, surface or over the entire volume; by moment of concentrated forces etc.

Generally, the similitude theory does not demand a particular relationship to exist between two identical loads on two similar structures, as determined by the modulus of geometric similarity (linear scale). However, the relationship between elements of deformation of such structures would evidently depend on the nature of loads applied at identical points in the structures. For example, the struts should be fixed in an identically equal manner (kinematically speaking) both in the model and the full-scale. Let us illustrate this by a simple example of deformations in identical structures under their own weight, i.e., subjected to load proportional to the cube of linear scale

$$G = \gamma l_c^3. \qquad \ldots (2.42)$$

Here, similar structures mean those fabricated from one and the same material and conforming to the law of geometric similarity with a certain linear scale l_c.

When such structures are deformed by an external force, they are subjected to forces whose ratios to each other are proportional, to some extent, to the linear scale l_c. The power index of l_c is obtained from the ratio of external load to the applied load expressed through the linear scale in the form of formula (2.42). Thus, at the assigned load determined by (2.42), the structures are subjected to the action of forces proportional to the linear scale l_c in the first power for forces distributed over the areas; proportional to the linear scale l_c in the second degree for forces distributed along the length; and proportional to l_c in the third degree for summed up and concentrated forces.

This is explained through an example of two similar beams using the known relationships for forces and elements of deformation of the beams: shearing forces

$$Q_1 = \int_0^{x_1} q_1\,dx_1 + \Sigma P_1 = \int_0^{x_2} l_c^2 q_2\,d(l_c x_2) + l_c^3 \Sigma P_2 =$$

$$= l_c^3 \int_0^{x_2} q_2\,dx_2 + l_c^3 \Sigma P_2 = l_c^3 Q_2; \qquad \ldots (2.43)$$

bending moments

$$M_1 = \int_0^{x_1} Q_1\,dx_1 = \int_0^{x_2} Q_2 l_c^3 l_c\,dx_2 = l_c^4 \int_0^{x_2} Q_2\,dx_2 = l_c^4 M_2 \qquad \ldots (2.44)$$

Normal stresses

$$\sigma_1 = M_1/W_1 = l_c^4 M_2/(l_c^3 W_2) = l_c M_2/W_2 = l_c \sigma_2. \qquad \ldots (2.45)$$

Curvature of bending line of beam

$$y_1'' = M_1/(EI_1) = l_c^4 M_2/(l_c^4 EI_2) = M_2/(EI_2) = y_2''. \qquad \ldots (2.46)$$

Angles of rotation of sections with reference to normal stresses

$$y_1' = \int_0^{x_1} y_1''\,dx_1 = \int_0^{x_2} y_2'' l_c\,dx_2 = l_c \int_0^{x_2} y_2''\,dx_2 = l_c y_2'. \qquad \ldots (2.47)$$

Deflection of beam

$$y_1' = \int_0^{x_1} y_1'\,dx_1 = \int_0^{x_2} l_c y_2' l_c\,dx_2 = l_c^2 \int_0^{x_2} y_2'\,dx_2 = l_c^2 y_2. \qquad \ldots (2.48)$$

Potential energy in beam

$$V_1 = \int_0^{x_1} \frac{M_1^2}{2EI_1}\,dx_1 = \int_0^{x_2} \frac{l_c^8 M_2^2}{2EI_c^4 I_2} l_c\,dx_2 =$$

$$= l_c^5 \int_0^{x_2} \frac{M_2^2}{2EI_2}\,dx_2 = l_c^5 V_2. \qquad \ldots (2.49)$$

It can be seen from this example that under the action of their own weight similar structures would have total displacements proportional to l_c^2, while the stresses would be proportional to l_c.

If the external loads (concentrated loads and overall distributed loads) are proportional to the square of linear scale, then the intensity of forces distributed lengthwise would be proportional to l_c in the first degree, while the intensity of forces distributed over the area remains unchanged; in other words, it is independent of the scale of the structures. Then, relations (2.42) to (2.48) for similar beams are written in this form:

$$Q_1 = l_c^2 Q_2, \quad M_1 = l_c^3 M_2, \quad \sigma_1 = \sigma_2, \quad \tau_1 = \tau_2,$$
$$y_1'' = y_2''/l_c, \quad y_1' = y_2', \quad y_1 = l_c y_2, \quad V_1 = l_c^4 V_2. \qquad \ldots (2.50)$$

Loads causing loss of stability in elastic systems completely satisfy these conditions. Hence in the type of modelling presently considered, not only the safety factor of structures based on stresses σ and τ would remain the same, but also the stability of structures based on the corresponding loads and stresses, for example:

a) Euler concentrated load for a beam under compression:

$$P_{e_1} = K\pi^2 EI_1/l_1^2 = K\pi^2 E l_c^4 I_2/(l_c^2 l_2^2) = l_c^2 P_{e_2}$$

and the corresponding stress

$$\sigma_{e_1} = P_{e_1}/S_1 = l_c^2 P_{e_2}/(l_c^4 S_2) = P_{e_1}/S_2 = \sigma_{e_2}.$$

b) Euler distributed load along length when beam is subjected to reduction:

$$q_{e_1} = 3EI_1/r_1^3 = 3El_c^4 I_2/(l_c^3 r_2^3) = l_c 3EI_2/r_2^3 = l_c q_{e_2}$$

and the corresponding stress

$$\sigma_{e_1} = q_{e_1} r_1/S_1 = l_c q_{e_2} l_c r_2/(l_c^2 S_2) = q_{e_2} r_2/S_2 = \sigma_{e_2}.$$

In general, if external loads on two similar structures are defined by a certain arbitrary function of linear scale $f(l_c)$, the corresponding relationships between other types of loads or forces in the structure would be as follows:

for forces distributed along the length $\frac{1}{l_c} f(l_c)$;

for forces distributed over the area $\frac{1}{l_c^2} f(l_c)$;

for forces distributed over the volume $\frac{1}{l_c^3} f(l_c)$;

for concentrated forces $f(l_c)$.

Then the load and deformations in the beams are expressed by the formulae:

$$Q = f(l_c) Q_2, \qquad M_1 = l_c f(l_c) M_2, \qquad \sigma_1 = \frac{1}{l_c} f(l_c) \sigma_2,$$

$$y_1'' = \frac{f(l_c)}{l_c^3} y_2'', \qquad y_1' = \frac{f(l_c)}{l_c^2} y_2', \qquad y_1 = \frac{f(l_c)}{l_c} y_2,$$

$$V_1 = \frac{f^2(l_c)}{l_c} V_2.$$

Let us now consider modelling equilibrium in elastic structures built from different materials.

Obviously, the elastic properties of materials—modulus of elasticity E and Poisson ratio ν—should be introduced as the governing parameters. To determine all the dimensions of a model, it suffices if one of its dimensions B is assigned (in doing so, it is understood that geometric similarity is observed). If apart from the applied loads P, the weight of the structure is also significant, then its density $\gamma = \rho g$ should also be taken into account. The system of governing parameters would thus be

$$\nu, E, B, \rho, g, P. \quad \ldots (2.51)$$

Using the π-theorem, three similarity criteria may be obtained from among the above parameters:

$$\nu, E/(\rho g B), P/(EB^2). \quad \ldots (2.52)$$

At $E_N = E_M$, it follows from the second criterion, that it is satisfied if $g_c = 1/(\rho_c B_c)$, which can be achieved in practice by means of a centrifuge.

From the third criterion it follows that similarity in its strictest sense exists only while evaluating equilibrium when the magnitudes of loads are proportional to the square of modulus of geometric similarity. This agrees entirely with the earlier mentioned relative influence of weight of structure on the deformation of its elements.

In conclusion, let us consider the question of modelling deformations in structures, using small-scale structures.

Let us denote by l the variation in length due to deformation of a certain element in an elastic system. The governing parameters and the similitude criteria in this case would be equal to the quantities expressed by (2.51) and (2.52) just as in the previous case.

Then for relative elongation, the relationship may be written as:

$$l/B = f\left(\nu, \rho g B/E, P/(EB^2)\right).$$

From this it is clear that l/B decreases as the density of structural elements is reduced, i.e., with reduction in $\rho g B/E$. Now, let us consider structures of different dimensions made from the same material, i.e., ν and E are the same. Let us also assume that external loads vary according to the third criterion, i.e., proportional to the square of dimensions: $P = EB^2$.

In such a case the magnitude of the criterion $\rho g B/E$ decreases with a reduction in structural dimensions; in other words, mechanical similitude is violated. In structures of smaller dimensions the relative deformations are also lower; hence such structures would be stronger. But such a conclusion is justified only when the density of material $\gamma = \rho g$ plays a significant role. If the weight of material is not significant, but $P/(EB^2) = \text{const}$, then the values of relative deformations would be the same for bodies of different scales.

The case of strength build-up when the structure's dimensions are diminished may even be illustrated when weight is not a significant factor. This occurs when the external load is proportional to the cube of linear dimensions. Here, l/B for smaller structures would be small compared to that of larger ones, and the external load would drop as the dimensions of the structure are reduced. Consequently, in this case the strength of structure increases with a reduction in its dimensions.

2.4 Approximate Evaluation of Pressure Fields in Blasts under Water and in Air in an Unconfined Medium

The aforementioned laws of similitude theory in blasts in air and under water have been validated by wide-ranging experimental investigations on shock waves. This enabled development of simple computational formulae convenient for practical use, as shown below. We can examine these first.

2.4.1 Pressure Field Induced by Blasting in Air

Using the general functional relationship (2.32)

$$\Delta p_f = C_4 f(\sqrt[3]{G}/r),$$

where G is charge weight and r distance from the blast centre. Through a comprehensive analysis of experimental data obtained from within the country and abroad, M.A. Sadovsky established a relationship for the excess pressure at the shock front generated by blasting a concentrated standard Trotyl charge (see subsection 2.1.2) in air:

$$\Delta p_f = (74.5\sqrt[3]{G}/r + 250\sqrt{G}/r^2 + 637G/r^3)10^3, \qquad \ldots (2.53)$$

where $\Delta p_f = p_f - p_0$ is the excess pressure at the shock front, Pa; G the charge weight, kg; r distance from the blast centre, m; p_f pressure at wavefront; p_0 pressure in undisturbed air medium.

If the charge weight G is expressed in terms of radius of equivalent spherical charge according to (2.30), formula (2.53) may be written as:

$$\Delta p_f = \left[(1.403/\bar{r} + 88.9/\bar{r}^2 + 4280/\bar{r}^3)\right] 10^6, \qquad \ldots (2.54)$$

where $\bar{r} = r/R^*_{0_z}$.

To obtain the maximum pressure in blasts of other *EC*, relationships (2.53) and (2.54) may be used by inserting in them the corresponding weight equivalent to the Trotyl charge G^* or radius $R^*_{0_z}$ of charge which, according to the energy law of similitude, is calculated by formulae (2.28) and (2.31), where $Q_2 = Q_{EC}$; $Q_1 = Q_{TP}$ the heat generated by blast conversion of charge used and Trotyl charge and G_{EC} is the actual weight of the explosive charge used.

The period of action of the positive phase of the pressure-time curve for a Trotyl charge is characterised by the relationships (τ_+, s)

$$\tau_+ = 1.5\sqrt[6]{G}\sqrt{\bar{r}} \times 10^{-3}, \qquad \ldots (2.55)$$

or in terms of charge radius

$$\tau_+ = 6.5R_{0_z}\sqrt{\bar{r}} \times 10^{-3}. \qquad \ldots (2.56)$$

From formulae (2.55) and (2.56), directly as the shock wave advances, duration of the compressive phase increases. This fact is also well explained by physical considerations, in particular by the fact that a wavefront moves with velocity $N_f > a_0$ (where a_0 is velocity of sound) while its 'tail', in which the excess pressure is nearly equal to zero, moves slowly with velocity of sound a_0, as a consequence of which the wave becomes stretched in time. However, the spatial length of wave (wavelength) remains almost unchanged as the duration of action increases concomitant with a drop in velocity N_f. The small increments in wavelength observed in this case can be estimated, according to M.A. Sadovsky, by the approximate formula

$$\lambda = \lambda_0 + a \ln r, \qquad \ldots (2.57)$$

where λ—wavelength; a—constant.

M.A. Sadovsky also showed that over a very wide range of distances, the pressure pulse in the compression zone may be determined by the following approximate relationship:

$$I_+ = \int_0^{\tau_+} p dt = AG^{2/3}/r = BR_{0_z}/\bar{r}, \qquad \ldots (2.58)$$

where I_+ is the pulse in the compression zone, MPa·s; A, B are constant coefficients equal to the approximate values 3.33–3.53 and (1.18–1.26) 10^3 respectively for Trotyl.

The wave pulse in the rarefied phase is determined by the empirical formula

$$I_- = I_+(1 - 1/(2r)), \qquad \ldots (2.59)$$

from which it can be seen that as the wave advances (increase of r), the value of I_- approaches that of I_+.

The nature of pressure variation over time at a fixed distance r depends on pressure at the wavefront. This is clearly seen from Fig. 5 in which a series of curves $\Delta p/\Delta p_f$ is represented for a wide range of Δp_f as a function of non-dimensional time t/τ_+.

A part of the curves may be approximated for $\Delta p_f < 10^5$ Pa by the following relationship:

$$\Delta p(t) = \Delta p_f(1 - t/\tau_+). \qquad \ldots (2.60)$$

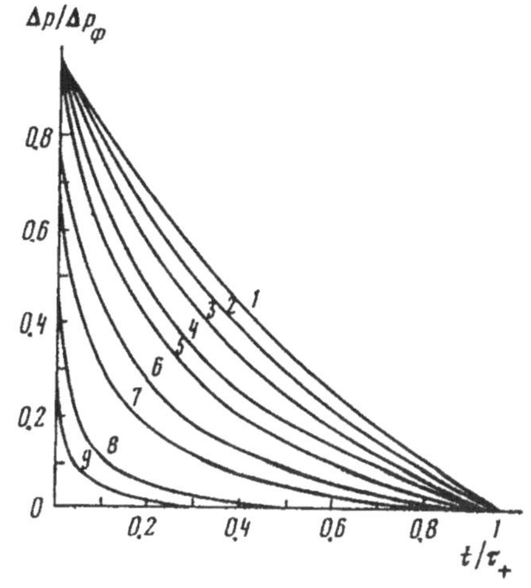

Fig. 5. Curves of pressure variation $\Delta p/\Delta p_f$ over time t/τ_+ for air shock waves: 1—9 at corresponding pressure intensities at wavefront Δp_f = 0.02, 0.05, 0.1, 0.2, 0.3, 0.5, 1, 5 and 20 MPa.

All these formulae for evaluating τ_+ and $\Delta p(t)$ are valid for blasting a Trotyl charge. To determine τ_+ and to plot the curves $\Delta p(t)$ in blasts of non-Trotyl charges, these formulae may be used by substituting in them corresponding values of equivalent Trotyl-based weight G^* or radius $R^*_{0_z}$ of the other charge. Then the pulse in compression phase is determined by numerical integration of the curve, plotted either on the basis of derived formulae and curves (see Fig. 5) or on the relationships:

$$I_+ = \int_0^{\tau_+} \Delta p(t)\,dt = \mu_i S; \qquad \ldots (2.61)$$

$$\mu_i = \Delta p_f t_M / A_M, \qquad \ldots (2.62)$$

where μ_i is the dimensional scale coefficient, Pa·s/mm^2;
Δp_f maximum pressure at the wavefront, Pa;
A_M magnitude of Δp_f in the oscillogram, mm;
t_M time scale in the oscillogram, s/mm;
S area of diagram showing compression zone of wave in the oscillogram, mm^2.

It is often necessary to know the specific energy or density of wave energy in order to characterise the effect of an air blast. This is determined by the formula

$$E_0 = \int_0^{\tau_+} \Delta p(t) v(t) dt = \frac{1}{\rho_0 a_0} \int_0^{\tau_+} \Delta p^2(t) dt.$$

A diagram of the compressive phase of a wave or the magnitude of pulse in the compression phase may also be used for computing specific energy. The formula in this case would be:

$$E_0 \simeq \mu_e S, \quad \ldots (2.63)$$

$$\mu_e = \frac{0.2 \Delta p_f^2 t_M}{\rho_0 a_0 A_M} = \frac{0.2 \Delta p_f}{\rho_0 a_0} \mu_i, \quad \ldots (2.64)$$

$$E_0 = \frac{0.2 \Delta p_f}{\rho_0 a_0} \mu_i S = \frac{0.2 \Delta p_f}{\rho_0 a_0} I, \quad \ldots (2.65)$$

where μ_e is the dimensional scale coefficient, $J/m^2 \cdot mm^2$; E_0 the specific energy in the compression wave phase, J/m^2.

A rock massif may be assumed to be an absolute rigid barrier compared to air while evaluating the pressure fields in ground blasts, which are almost akin to those in large-scale blasts in open-pit mines. In such a case it may be assumed that the major part of the blast energy will be distributed in aerial semi-space. From this point of view, a blast near the earth's surface happens to be equivalent to a blast of charge weighing two times more in an unconfined atmosphere. Formulae (2.53) and (2.54) will remain valid if two times the charge weight $2G$ is substituted in them. Then for the true weight G, formula (2.53) is written as:

$$\Delta p_f = (93.2\sqrt[3]{G}/r + 383\sqrt[2]{G}/r^2 + 1470G/r^3)10^3, \quad \ldots (2.66)$$

or in terms of charge radius

$$\Delta p_f = (1.75/\bar{r} + 136/\bar{r^2} + 8550/\bar{r^3})10^6. \quad \ldots (2.67)$$

The duration of positive wave phase and pulse would be respectively equal to:

$$\tau_+ = 1.7 \times \sqrt[6]{G}\sqrt{\bar{r}}10^{-3}; \quad \ldots (2.68)$$

$$\tau_+ = 7 \times R_{0_z}\sqrt{\bar{r}}10^{-3}; \quad \ldots (2.69)$$

$$I_+ = (5.1 \text{ to } 5.5)G^{2/3}/r = (1810 \text{ to } 1950)R_{0_z}/\bar{r}. \quad \ldots (2.70)$$

In this case the pressure diagram and energy calculation are done using the formulae applicable for unconfined atmosphere.

2.4.2 Pressure Field Induced by an Underwater Explosion

On the basis of a large number of experiments, R. Kaul established the following relationship for pressure (MPa) at the wavefront P_m generated by firing a standard Trotyl charge:

$$p_m = 52.3(\sqrt[3]{G}/r)^{1.13}. \qquad \ldots (2.71)$$

This relationship for non-dimensional distances $\bar{r} = r/R_{0_z}$ would be of the form

$$p_m = 1440\gamma_i/\bar{r}^{1.13}, \qquad \ldots (2.72)$$

where γ_i is the transformation coefficient for a non-Trotyl charge

$$\gamma_i = (\rho_{EC}Q_{EC}/(\rho_{TR}Q_{TR}))^{\frac{1.13}{3}} = (\rho_{EC}Q_{EC}/(\rho_{TR}Q_{TR}))^{0.377}, \qquad \ldots (2.73)$$

where ρ_{EC}, ρ_{TR} are density of charge of *EC* and Trotyl respectively; Q_{EC}, Q_{TR} the heat generated in transformation of charge due to blasting of *EC* and Trotyl respectively.

These formulae are justified for relative distances $\bar{r}$ in the interval $12 \leqslant \bar{r} \leqslant 240$.

For still shorter distances a more accurate relationship was derived by B.V. Zamishlyaev and Yu.S. Yakovlev:

$$p_m = 44.1\left(G^{1/3}/r\right)^{1.5} = 3610/\bar{r}^{1.5}, \qquad \ldots (2.74)$$

valid for relative distances $6 < \bar{r} \leqslant 12$.

The time of entry of the shock front from the moment of firing or its relative value is given by

$$\bar{t}_f = t_f a_0/R_{0_z},$$

where a_0 is the velocity of sound in water ($a_0 = 1460$ m/s); for Trotyl charges it may be determined for the relative distances $\bar{r} = 2$ to 12.

$\bar{r}$ ...	5	6	7	8	10	12
$\bar{t}_f$ ...	2.3	3.1	3.9	4.6	6.4	8.2

The diagram of pressure in the positive phase of a wave is described by the following relationships:

$$p(t) = p_m e^{-\frac{1}{\theta}\left(t-\frac{r}{a_0}\right)}\varepsilon\left(t-\frac{r}{a_0}\right);$$

$$\theta = (G^{1/3}/r)^{-0.24}G^{1/3}10^{-4},$$

or

$$\theta = 1.4\frac{R_{0_z}}{a_0}\bar{r}^{-0.24}.$$

According to the data of B.V. Zamishlyaev and Yu.S. Yakolev, the diagram of a shock wave may be computed at $p_0 < 10^6$ Pa based on the following relationships:

$$p(t) = p_m \left\{ \begin{array}{ll} e^{-t/\theta} & \text{for } t \leqslant \theta; \\ \dfrac{0.368}{t/\theta} & \text{for } \theta < t < (5 \text{ to } 10)\theta, \end{array} \right\} \qquad \ldots (2.75)$$

where θ is a constant of exponential drop (c), determined by the formula

$$\theta_{a_0}/R_{0_z} = 3.5\sqrt{\log \bar{r} - 0.9}, \quad \text{for } 30 < \bar{r} < 240.$$

The time of action of the positive phase of the shock wave in water is $\tau_+ \approx 20R_{0_z}/a_0$.

When formula (2.75) is used, the positive phase pulse of a shock wave would be:

$$I_+ = \int_0^{\tau_+} p(t)dt = p_m\theta \begin{cases} 1 - e^{-t/\theta} & \text{at } t < \theta \\ 10.632 + 0.368 \ln \frac{t}{\theta} & \text{at } \theta < t < 10\theta. \end{cases} \quad \ldots (2.76)$$

When formula (2.76) is used, the shock-wave pulse is given by

$$I_+ = \int_0^{\tau_+} p(t)dt = p_m\theta \left[1 - e^{-1/\theta(t-r/a_0)}\right] \varepsilon(t - r/a_0).$$

Density of energy flux of a shock wave is determined by the relationship

$$E_0 = \int_0^t \rho v \Delta(u + v^2/2 + p/\rho)dt.$$

For pressures $p < 100$ MPa, this may be approximately calculated by the formula:

$$E_0 = \left[p_m^2\theta/(2\rho_0 a_0)\right] (C + Da_0\theta/2), \quad \ldots (2.77)$$

where $C = \begin{cases} 1 - e^{-2t/\theta} & \text{for } t < \theta, \\ (1.135 - 0.27(t/\theta)) & \text{for } \theta < t < 10\theta; \end{cases}$

$$D = \begin{cases} 1 - e^{-2t/\theta}\,(2 - e^{-t/\theta}) & \text{for } t \leqslant \theta; \\ (0.4 + 0.135 \ln(2t/\theta)) & \text{for } \theta < t \leqslant 10\theta. \end{cases}$$

Formulae (2.71) to (2.77) are derived for a Trotyl charge of density 1.6×10^3 kg/m^3 and heat of blast transformation $Q_{TR} = 4440$ kJ/kg. To use them for other explosives, it is necessary to introduce into them the equivalent Trotyl weight of the other explosive charge weight G^* or the other explosive charge radius $R^*_{0_z}$, as determined by corresponding formulae (2.28) and (2.31). In doing so, the pulses and energy of the compression phase should be computed according to (2.76) and (2.77).

2.5 Similitude Criteria Applied to Blasting in Rocks

Let us first study the features of the phenomena accompanying a blast in rocks. The modern concept on the development of a blast in rock is summarised below.

When a charge of *EC* is detonated, a detonation wave with a shock front forms, entrailing a zone of reaction and field of expanding gaseous blast products. The detonation wave and gaseous products, on reaching the boundary of separation, heavily compress the medium surrounding the charge. As a result a wave of compression is generated in the rock, which advances symmetrically with reference to the charge. Further, a zone of crushed and stressed rock also forms near the charge. Since the energy of the compression wave dissipates faster during its propagation, and due to the nature of the relationship between deformation and stresses in rocks, the compression wave per se loses the character of a shock wave and transforms initially into an elastoplastic and later into an elastic wave with a non-shock front. Actually, as can be seen from Fig. 6,a, when a detonation wave is refracted in rocks, a shock wave with a maximum pressure p_{ref} is also formed in it. Here the velocity of wave propagation N, proportional to the angle of inclination α_{ref} of secant curve of rock loading $\sigma = f(\varepsilon)$ drawn from the origin would be more than the velocity of sound in rock $c_0 = c_1 \tan \alpha_0$. Since the refracted shock wave proceeds outwards in rock, owing to dissipation of wave energy, its amplitude p_{ref} and, correspondingly, velocity of propagation N, would decay and attain at the end of section 1 the values $p = \sigma_{ref}$ and $N = c_0$. Thus in section 1 the shock front of the wave remains unaffected. Further advancement of the wave in section 2-3 results in its gradual transformation into a non-shock one (Fig. 6, b), owing to the fact that rock particles through which the section of wave with maximum amplitudes σ_1 and σ_2 passes, would move at lower velocities $c = c_1 \tan \alpha$ than those rock particles through which the wave with low amplitudes $\sigma < \sigma_e$ passes. These rock particles move with a velocity of sound $c_0 \simeq \tan \alpha_0 > c \simeq \tan \alpha$ (see Fig. 6, a). During further propagation in section 4 this wave finally attains the form of stress wave with a non-shock front and behaves as an elastoplastic stress wave right up to the moment of decay in maximum amplitude to the stress level equivalent to the elasticity limit σ_e. During subsequent propagation in section 5 the wave converts into an elastic one which, at large distances and after a series of reflections and refractions from different layers of rock and soil, forms a wave train of every conceivable vibration; it is then termed a seismoblast or, simply, a seismic wave. Such waves are differentiated here into straight, refracted and reflected body waves (longitudinal and transverse) as well as different types of surface waves (Love, Rayleigh etc).

Gaseous products of explosion expanding further compress the crushed rock mass and form a cavity when the free faces are located quite some distance away from the charge. If the explosion occurs in the vicinity of free faces and the energy of gaseous products is adequate enough to counteract the throw of broken rock, a blasting cone forms. At this juncture a specific role is played by the tensile wave reflected from the free face, which causes

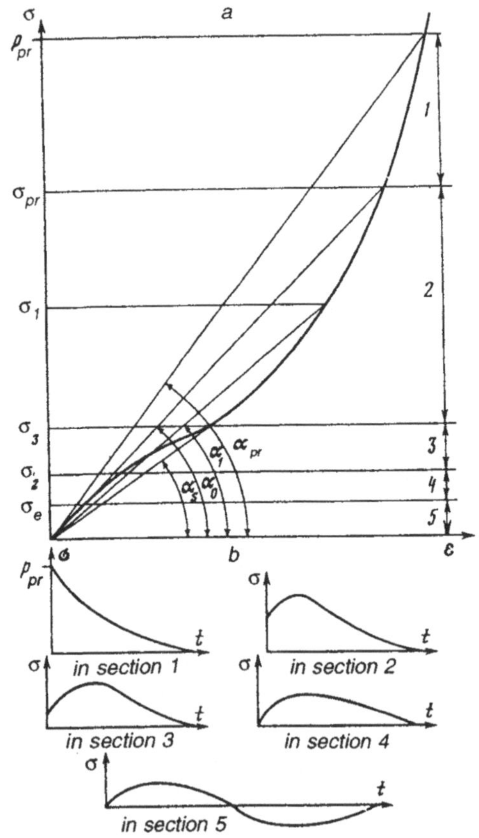

Fig. 6. a: Curve of dynamic compression of rocks; b: diagram of stress wave during propagation in rock.

slabbing of rock at the free face, opening of cracks in the conical region of throw when it proceeds deep into the rock to the blast centre and separation of broken rock at the fracture formed between charges.

Such a picture is characteristic for blasting of concentrated (spherical) charges. When an extended (cylindrical) charge is fired, the wave picture becomes somewhat complicated. In this case besides the compressional wave, a shear wave is also generated as a result of propagation of the detonation wave along the blasthole wall towards the charge axis (Fig. 7). However, the magnitude of this wave is much less than that of the compressional wave due to the limited resistance of rock to forces directed along the blasthole walls, given the low cohesive forces of *EC* and rock in this direction. Therefore, in most cases its effect is neglected to simplify the task and only the compressional wave is considered, which happens to be the result of rock deformation in a direction perpendicular to the wall of the blast hole. Actually, it may influence the overall wave picture by interfering non-linearly with the compressional wave, weakening its intensity and varying the form of the wave front with distance. However, the form of the compressional

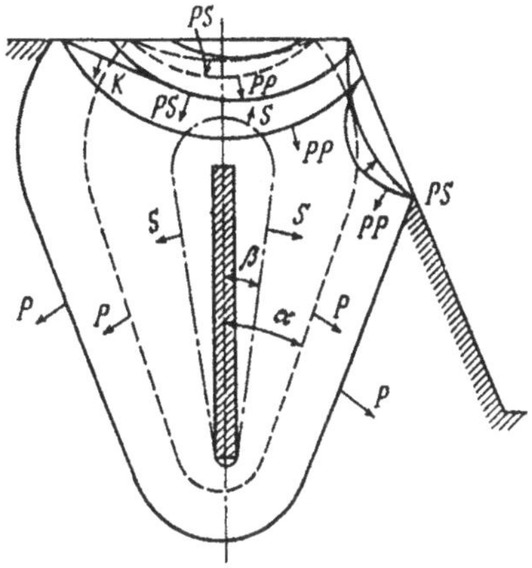

Fig. 7. Wave patterns when an extended charge is blasted in rock.
P—longitudinal wavefront; *S*—transverse wavefront; *PP* and *PS*—longitudinal and transverse waves respectively reflected from free face when a longitudinal wave *P* is incident on it; *K*—conical wave.

wave front may per se have a spherical surface, at a velocity of detonation wave D less than the velocity of propagation of compressional wave c_P, i.e., at $D < c_P$ (see Fig. 8, b), and may even have a pear-shaped surface at $D > c_P$, which happens to be a combination of surface of a cone truncated by hemispheres at its end (Fig. 8, a). The picture of waves reflected from the free face accordingly becomes more complex, especially in breakage of rocks in benches where two free faces exist: face and floor of the bench. As can be seen from Fig. 8, in this case when compressional waves *P* fall on the free bench face and bench floor, reflected longitudinal *PP* and transverse *PS* waves are generated which, in turn, mutually interact and create an even more complicated wave picture.

Two processes can be distinguished when a blast occurs in rock: Generation of a shock or compressional wave, causing initiation and opening up of existing cracks and local action of the blast, by which is meant final formation of the rock breakage zone and formation of a cavity or throw of broken rock due to the effect of expanding gaseous products of the blast. Here it is generally assumed that the major part of blast energy in rocks is expended in fragmenting and heating the neighbouring medium, while a small part remains within the gaseous products of the blast; the energy of seismic waves does not exceed 10% of the total blast energy [1]. Dissipation of blast energy in rocks, as distinct from that under water and in air, is determined largely by the strength characteristic of the medium σ_*, which is usually three times lower than the rigidity of the medium, rather than by the magnitude of rigidity of the medium $\rho_p c_p^2$ (where ρ_p density and c_p velocity

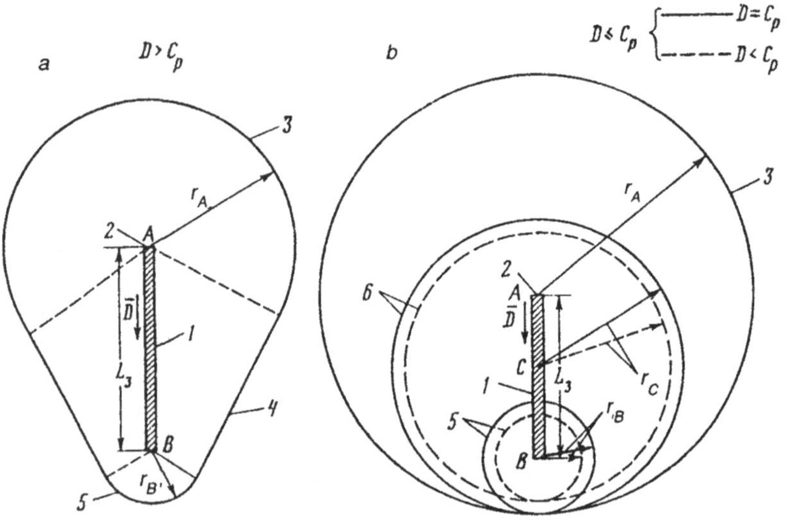

Fig. 8. Form of stress wavefronts created by blasting of an extended charge. 1—charge; 2—detonator; 3—spherical wavefront with centre at point A of charge initiation; 4—conical wavefront; 5—spherical wavefront generated by initiating charge at point B; 6—spherical wavefront generated by initiating charge at point C.

of sound in the medium). Therefore the zone of intensive dissipation of blast energy in rock would at least exceed by one order the zone of dissipation of a powerful shock wave under water and in air, because the radius of this zone would be determined by the quantity $r_* = (E_0/\sigma_*)^{1/3}$ and not by $r_0 = [E_0/(\rho_p c_p^2)]^{1/3}$, where E_0 is the energy released by firing a charge of *EC*. Moreover, as evidenced by experience in studying shock wave adiabatic expansion of various solid bodies and rocks, a compressional wave with shock front in rocks may be realised only at pressures of the order 10^4 MPa. This implies that the dimensions of the zone in which a compressional wave front exerts a shock impact are very limited and for chemical *EC* in hard rocks this zone does not exceed 2 to 3 times the charge radius.

The aforementioned features of a blast in rock, as distinct from that under water and in air, considerably complicate the task of modelling all phenomena occurring during rock blasting. As of today, general differential (integral-differential) equations are not available for describing concomitantly the entire set of phenomena associated with a rock blast.They are quite difficult to derive as many phenomena within the near-field zone of a blast have not been studied in a comprehensive manner. Hence the similitude criteria are derived from the dimension theory using the governing parameters and generalised co-ordinates that characterise the said phenomena of blast in rocks.

Obviously, the quantities characterising the charge, explosive column, conditions of blasting, the medium surrounding the charge-rock and also the spatial co-ordinates, time and acceleration due to gravity g should be included in the governing parameters. The parameters of stress wave and dimensions of breakage zones are attributed to generalised co-ordinates.

The quantities characterising a charge are mainly density ρ_{EC} or density of charging γ_{EC} equivalent to weight G_* or radius $R^*_{0_z}$ based on blast energy (energy E_0 or heat of blast transformation Q_{EC} of a charge). The explosive column is characterised by its dimensions: radius R_K of a spherical cavity or radius R_{bh} and length l_{bh} of blast hole etc. The conditions of blasting are indicated by the depth of charge placement H or line of least resistance W, length of stemming l_{stem} and type of stemming material, method of initiation etc. The quantities that characterise the medium surrounding the charge are density of medium ρ_p, velocity of propagation of longitudinal waves c_P and transverse waves c_S, modulus of elasticity E, Poisson ratio ν, critical stresses under compression σ_{compr}, under shear σ_{shear} and during rupture σ_{rup}, magnitude of relative ultimate deformations in volume ε_* and shear deformations γ_*, values relating to jointing of rock (average number of cracks n, average length l_{TR}, width b_{TR} and depth h_{TR} of cracks).

The stress-wave parameters would obviously be determined by amplitude of stress σ, particle (mass) velocity v, displacement of particles in the medium u, time-related parameters (time of growth of amplitudes to the maximum τ_N, duration of positive τ_+ and negative τ_- phases, period of vibration T), and also by certain other parameters describing the shape of the stress-wave diagram. The radius of cavity R_0 (in camouflet blast) or radius R_B and depth of cone H_B (in throw blast) and also radii of zone of compression R_{compr}, fragmentation R_{fr} and zone of crack-formation R_{TR} are attributed to the dimensions of fragmentation zones.

All the aforementioned parameters of stress and disintegration wave would depend on the governing parameters, from which a certain functional relationship can be formulated.

$$P_i = f_i(R^*_{0_z}, R_K, R_{bh}, l_{bh}, H, W, l_{stem}, \rho_{EC}, \rho_0, c_P, c_S, E, \nu,$$
$$\sigma_{compr}, \sigma_{shear}, \sigma_{rup}, \varepsilon_*, \gamma_*, l_{TR}, b_{TR}, h_{TR}, n, g, t, r(x,y,z)).$$

Let us transform this relationship, based on the π-theorem, into a relationship between non-dimensional combination of parameters or similitude criteria by taking, for example, the charge radius R_{0_z}, modulus of medium elasticity E and acceleration due to gravity g as the key quantities. The following criterial relationship is obtained after transformation:

$$P_i = f_i(R_K/R^*_{0_z}, R_{bh}/R^*_{0_z}, R_B/R^*_{0_z}, l_{bh}/R^*_{0_z}, H/R^*_{0_z},$$

$$W/R^*_{0_z}, l_{stem}/R^*_{0_z}, l_{TR}/R^*_{0_z}, b_{TR}/R^*_{0_z}, r/R^*_{0_z},$$
$$\rho_{EC}R^*_{0_z}g/E, \rho_0 R^*_{0_z}g/E, \sigma_{compr}/E, \sigma_{shear}/E, \sigma_{rup}/E,$$
$$c_p^2/(R^*_{0_z}g), c_S^2/(R^*_{0_z}g), t^2g/R^*_{0_z}, \nu, \varepsilon_*, \gamma_*, n).$$

In the above expression the first ten simplexes are the criteria arising out of conditions of geometric similitude with typical linear dimension equal to the energy-based equivalent charge radius. According to these criteria, all linear quantities depicting the studied phenomena of blast and attributed to the corresponding charge radii, should be equal for the model and the full-scale.

The subsequent five simplexes of the type $\rho_0 R^*_{0_z}g/E$ and σ_*/E happen to be similitude criteria, relating elastic and strength characteristics of the medium (σ_*) to inertial forces. These criteria are readily reduced to the Cauchy similitude criteria valid for elastic vibrations. Actually, if these criteria were divided by relative strain ε_0 (non-dimensional quantity), we would have

$$\frac{\rho_0 R^*_{0_z}g}{(E\varepsilon_*)} \to \frac{\rho l^2/t^2}{E\varepsilon_*} = \frac{\rho v^2}{E\varepsilon_*} = \frac{v^2}{\frac{\varepsilon_*}{\rho}E} \to \frac{v}{\sqrt{\frac{\varepsilon^*}{\rho}E}} = Ca;$$

$$\frac{\sigma_*}{E\varepsilon_*} \to \frac{F_*}{l^2E\varepsilon_*} = \frac{ml/t^2}{l^2E\varepsilon_*} = \frac{\rho l^4/t^2}{l^2E\varepsilon_*} = \frac{\rho v^2}{E\varepsilon_*} \to \frac{v}{\sqrt{\varepsilon_*E/\rho}} = Ca.$$

From these criteria, the criterion of dynamic similitude (Euler number) can also be obtained by dividing $\rho g R^*_{0_z}/E$ by σ/E and later multiplying by the subsequent criterion $c_P^2/(R^*_{0_z}g)$ or $v^2/(R^*_{0_z}g)$, finally obtaining $\sigma/(\rho v^2) = \ni$.

The complexes in which velocities of propagation $c_P, c_S, c_P^2/(R^*_{0_z}g)$ and $c_S^2/(R^*_{0_z}g)$ are contained are essentially the Froude similitude criteria relating inertial forces with gravitational ones. Actually

$$c_P^2/(R^*_{0_z}g) \to v^2/lg = Fr.$$

These criteria play a major role in the phenomena related to the effect of gravitational forces, for example, in the throw and displacement of rock mass, fly rock, heaping of such and also in the propagation of surface waves in large-scale blasts.

The last complex $t^2g/R^*_{0_z}$ is actually the criteria for kinematic similitude and reduces to the Strouhal number

$$t^2g/R^*_{0_z} \to t^2\frac{l}{t^2}/l = tv/l = H.$$

The parameters $\nu, \varepsilon_*, \gamma_*$ and n indicate that the Poisson ratio ν, magnitude of ultimate relative strains ε_* and γ_* and average number of cracks should be equal for both the medium being modelled and the actual medium used for simulation. Though the strength characteristics of these media might be different, but their ratios to the elasticity modulus σ_{compr}/E, σ_{shear}/E, σ_{dr}/E should be the same as those of the full-scale medium.

Analysing the dimensions of the governing parameters using the π-theorem, it can be established that for complete similarity of the processes occurring during a blast in rock, there should first of all exist geometric similarity of all linear dimensions with the scale equal to the ratio of charge radii of the full-scale and the model, which themselves are equivalents based on blast energy. Later, the kinematic and dynamic similitude or equality of the Strouhal and Euler numbers as well as equality of the Froude and Cauchy numbers characterising weight and elasticity respectively, are determined. Finally, the equality of Poisson ratios, ultimate relative strains, ratios of strength characteristics to the elasticity modulus and average number of fractures are established.

The entire set of these conditions (criteria) of similitude cannot be maintained simultaneously as difficulties are encountered while realising them in the model.

Thus in maintaining geometric similitude, problems of modelling the dimensions and average number of fractures arise due to the constraints of technical possibilities and also lack of knowledge of these parameters for the full-scale medium. This has always to be borne in mind while modelling fragmentation of rock mass, wherein natural joining plays a very important role.

Difficulties arise even in maintaining the equality of Froude number, i.e., while modelling the effect of gravitational force. In this case the modelling medium should usually be densified with the help of centrifugal forces (centrifugal modelling) or by hydrodynamic pressure or using a press or usual weights. In doing so, complications may arise due to the specific features of each of these methods, such as limited dimensions of the model (in the first two cases), need for considering the influence of weight of the press and the load (in the latter cases), and other peculiarities. Arising out of these, while investigating various phenomena of a blast in rocks, it is advisable to break the entire problem into a series of smaller problems, i.e., study the individual characteristic phenomena in isolation from others.

Thus while studying the laws of propagation of stress waves and their effect on the medium, it is necessary to insulate the local effect of a blast. The number of governing parameters reduce because the parameters for local effect as well as acceleration due to gravity are discarded. Only the similitude criteria characterising geometric and kinematic similarity, equality of strength and elastic properties of the media and Cauchy criterion remain for consideration. The most difficult part in this is the realisation of similitude of natural and induced jointing in the media, non-homogeneity of rock massif, i.e., the possibility of selecting or fabricating a model that simulates the massif while taking into account the geometric similarity. While modelling the phenomena of local effects of a blast, for example, the researcher can disregard the phenomena of wave propagation, but the body forces should be taken into account, such as parameter of acceleration due to gravity, g. This usually leads, as already mentioned, to centrifugal modelling using a centrifuge. In this case the number of governing parameters and the corresponding similitude criteria decrease but the strength properties of the media, Froude criterion and similarity of breakage in both the media remain. It is difficult to realise the last feature, especially when the same media are chosen for the model and the full-scale, because in such a case it is necessary to model granulometric composition of rock mass, forces resisting its movement and under complex hydrogeological conditions, for example while blasting in the direction of a waterlogged, mined-out area, to take into account the weight of water involved in motion.

2.6 Evaluation of Stress Field while Blasting in Rocks under Conditions of an Unconfined Medium

A large number of approximate empirical relationships differing in formulation and experimental coefficients associated with them, are available for calculating the parameters of stress waves. This is mainly because wide-ranging investigations in rocks, both monolithic and non-homogeneous, with the typical heterogeneities and varied jointing associated with them have been undertaken. A common feature in these relationships is the usage of extended geometric similitude (while formulating the law), directed at equivalence of governing parameters of stress (stress, velocity, displacement of particles) at corresponding spatial points $\bar{r} = r/R_{0_z}$ and at corresponding moments of time $\bar{t} = t/R_{0_z}$ or tc_p/R_{0_z} while considering the equivalent charge radii based on blast energy.

An entire set of such relationships exist for hard rocks, obtained on the basis of laboratory and full-scale experimental data collected by the staff of RGP Department of Leningrad Mining Institute, V.A. Borovikov, I.F. Vanyagin, A.N. Khanukaev and by the staff of other organisations —V.M. Gogolev, V.G. Myrkin, M.M. Pavlov, B.D. Kristoforov and others.

In a blast of a concentrated charge the curve of radial component of stress wave in the range of distances $r = (20 \text{ to } 100)R_{0_z}$ can be described by [3]

$$\sigma_2(r,t) = \sigma_{r_{max}}(r)e^{-\alpha(\tau-\tau_N)}\frac{\sin\beta\tau}{\sin\beta\tau_N}[\varepsilon_0(\tau) - \varepsilon_0(\tau-\tau_+)]. \qquad \dots (2.77)$$

The following quantities are introduced into this relationship. *Time elapsed from the moment of entry of stress wave at a given point r*, $\tau = t - t_{pr}$, where t is the time elapsed from the moment of firing a charge of *EC*; t_{ref} the time of entry of wave to the given point r from the moment of firing.

The maximum value of stress (Pa) in the wave at the given point r

$$\sigma_{r_{max}}(r) = \rho_p c_p \; 2.25 \times 10^3/\bar{r}^2, \qquad \dots (2.78)$$

where $\rho_p c_p$ is the acoustic rigidity of rock, N·f/m^3; $\bar{r} = r/R_{0_z}$ the relative distance.

Time (s) of growth of stress amplitude to its maximum value τ_N*:*

$$\tau_N = (a_1 + a_2\bar{r})R_{0_z} \qquad \dots (2.79)$$

Coefficient β, characterising the duration τ_+ of the positive phase of stress wave (1/s)

$$\beta = \pi/\tau_+ = (b_1 + b_2\bar{r})/R_{0_z}. \qquad \dots (2.80)$$

Coefficient α, characterising the rate of growth and decay of amplitude of stress wave over time (1/s)

$$\alpha = \beta \cot\,(\beta\tau_N). \qquad \dots (2.81)$$

In (2.79) and (2.80) the quantities a_1 (s/m), a_2 (s/m), b_1 (m/s), b_2 (m/s) are the numerical dimensional coefficients depending on the type of rock. They are determined by the expressions:

$$\begin{aligned} a_1 &= (0.02 + 0.87 \times 10^{-8}\rho_p c_p)10^{-3}, \\ a_2 &= (0.19 + 0.122 \times 10^{-7}\rho_p c_p)10^{-4}, \\ b_1 &= 150 + 27 \times 10^{-6}\rho_p c_p, \\ b_2 &= -0.2 \times 10^{-7}\rho_p c_p. \end{aligned} \qquad \dots (2.82)$$

The tangential component of stress wave curve may be computed by the following relationships [3]

$$\sigma_\varphi(r,t) = \sigma_{\varphi max}(r)\left\{\frac{1}{3}\sin\left(\frac{\pi\tau}{\tau_N}\right)[\varepsilon_0(\tau) - \varepsilon_0(\tau - \tau_N)] - \right.$$
$$\left. -e^{-\alpha(\tau - 2\tau_N)}\frac{\sin[\beta(\tau - \tau_N)]}{\sin(\beta\tau_N)}[\varepsilon_0(\tau - \tau_N) - \varepsilon_0(\tau - \tau_N - \tau_+)]\right\}, \qquad \dots (2.83)$$

$$\sigma_{\varphi max}(r) = \left[1 - 2\left(\frac{c_S}{c_P}\right)^2\right]\sigma_{r_{max}}(r) = \frac{\gamma}{1-\gamma}\sigma_{r_{max}}(r), \qquad \dots (2.84)$$

where c_S, c_P are velocity of propagation of transverse and longitudinal waves respectively in a given rock.

In the expressions derived above, $\varepsilon_0(\tau)$ is a unique discontinuous function of 0th order

$$\varepsilon_0(\tau) = \begin{cases} 1 & \text{at } \tau \geq 0 \\ 0 & \text{at } \tau < 0. \end{cases}$$

The mass velocity or velocity of rock particles is given by

$$v(r,t) = \sigma(r,t)/(\rho_p c_p).$$

The maximum displacement of rock particles in the stress wave is computed by integrating the curve of rock particle displacement versus time within the limits $r = 0$ to $\tau = \tau_+$, i.e.,

$$u_{r_{max}}(r) = \int_0^{\tau_+} v_r(r,t)\,dt.$$

This can be calculated approximately using the formula:

$$u_{r_{max}}(r) \approx \frac{1}{2} v_{r_{max}}(r)\tau_+.$$

In the case of blasting an extended charge relationships (2.77), (2.79) to (2.81) and (2.83) may also be approximately used, but the coefficients in formulae (2.82) would differ and may be calculated [4] by

$$a_1 = (0.325 + 0.161 \times 10^{-6}\rho_p c_p)10^{-3},$$
$$a_2 = (0.47 + 0.113 \times 10^{-7}\rho_p c_p)10^{-4},$$
$$b_1 = 178 + 3.49 \times 10^{-6}\rho_p c_p,$$
$$b_2 = -0.125 - 0.218 \times 10^{-7}\rho_p c_p.$$

The maximum amplitudes $\sigma_{r_{max}}$ and $\sigma_{\varphi_{max}}$ are calculated using the formulae

$$\sigma_{r_{max}}(r) = 545\rho_p c_p/\bar{r}^{1.1}; \qquad \ldots (2.85)$$

$$\sigma_{\varphi_{max}}(r) = (c_1 + c_2 r)\sigma_{r_{max}}(r), \qquad \ldots (2.86)$$

where c_1 and c_2 are dimensionless coefficients determined by

$$\left.\begin{aligned} c_1 &= 0.09 + 0.228 \times 10^{-7}\rho_p c_p, \\ c_2 &= -(0.07 - 0.224 \times 10^{-7}\rho_p c_p)10^{-2}. \end{aligned}\right\} \qquad \ldots (2.87)$$

It should be borne in mind that relationships (2.78) and (2.84) were obtained for blasting charges of Trotyl with a density of 1.6×10^3 kg/m^3 and heat of blast transformation 4440 kJ/kg. In the case of firing concentrated charges of other explosives, these relationships may be used but the charge radius of explosive charge equivalent to that of a Trotyl charge based on blast energy should be determined:

$$R^*_{0_z} = R_{0_z}\left(\frac{\rho_{EC} Q_{EC}}{\rho_{TR} Q_{TR}}\right)^{1/3}, \qquad \ldots (2.88)$$

where ρ_{EC}, Q_{EC} and ρ_{TR}, Q_{TR} are density and heat of blast transformation of the explosive used and that of Trotyl; R_{0_z} is the radius of charge containing the other explosive.

The equivalent charge radius thus obtained is introduced into the derived relationships while calculating the relative distances and all other quantities that determine the stress wave parameters.

The formulae derived for an extended charge should be treated similarly, i.e., with relationships (2.85) and (2.86) obtained for charges of TEN* with $\rho_T = 1.5 \times 10^3$ kg/m^3, $Q_T = 5860$ kJ/kg. While calculating according to these formulae, the radius of charge continuing the given explosive equivalent to that of charge of TEN based on blast energy should be determined:

$$R^*_{0_z} = R_{0_z}[\rho_{EC} Q_{EC}/(\rho_T Q_T)]^{1/2}, \qquad \ldots (2.89)$$

where ρ_T, Q_T and ρ_{EC}, Q_{EC} are density and heat of blast transformation of TEN, Hexagene** and other explosives being used; R_{0_z} the radius of charge containing the other explosive.

It needs to be highlighted that the relationships detailed above were obtained on an experimental theoretical basis [4] for the middle-field zone of a blast, i.e., for relative distances $\bar{r} = 20$ to 100, while for extended charges they are valid in areas parallel to the axis and situated roughly at the middle portion of these charges. At the remaining spatial points surrounding the extended charge, the maximum amplitudes would have different values and the shape of stress-versus-time curve would differ. For qualitative assessment of the effect of stress wave in rocks induced by firing extended charges, the derived relationships are valid and provide satisfactory results in predicting the effect of field blasts.

* Chemical formula: $C_5H_8(ONO_2)_4$ —Translator.

** Chemical formula: $C_3H_6N_6O_6$ —Translator.

3

Physical Modelling of Blast Effects in Rock Breakage

Physical modelling is currently widely used in scientific investigations for the manufacture and development of equipment. The theories of similitude and dimension provide the theoretical basis for such modelling. Physical modelling is particularly suited for studying the various phenomena associated with the process of rock breakage by blasting.

The *essence of physical modelling* lies in the method of investigation using physical (technical) models and in extrapolating results of such investigations to the full-scale (under study) phenomenon with a view to obtaining essential information about the nature of the processes occurring and the variation of different quantities associated with the said phenomenon in the full-scale. The physical model happens to be an object or a special device capable of replacing the full-scale object and its study ought to yield new information about the latter. In the simplest case a physical model replicates the phenomenon under study while preserving geometric similarity and the physical nature, differing from the full-scale only in dimensions and rate of occurrence of the process. In more complicated cases the model is fabricated from materials that differ from the full-scale, but adhere strictly to the conditions of similitude, i.e., the model is prepared from equivalent materials.

3.1 Essential Features of Physical Modelling Methods Based on Newton's Mechanical Similitude and Illustrations Related to their Use

3.1.1 Theoretical Principles of Mechanical Modelling Methods

Newton's theory of mechanical similitude is the basis for the modelling methods considered here. It presupposes geometric, kinematic and dynamic similitude:

$$l_{N_1}/l_{M_1} = l_{N_2}/l_{M_2} = \ldots = l_{N_n}/l_{M_n} = l_c,$$
$$t_{N_1}/t_{M_1} = t_{N_2}/t_{M_2} = \ldots = t_{H_n}/t_{M_n} = t_c,$$
$$m_{N_1}/m_{M_1} = m_{N_2}/m_{M_2} = \ldots = m_{N_n}/m_{M_n} = m_c,$$

where l_{N_1} and l_{M_1}; t_{N_1} and t_{M_1}; m_{N_1} and m_{M_1} are corresponding linear dimensions, time periods and masses of the full-scale and the model; l_c, t_c, m_c are geometric, time-related and mass-related scales of similitude.

It is common knowledge that by mechanical similitude according to Newton we mean the identity of similitude criteria of Newton in the full-scale and the model. This arises from Newton's second law and is of the type $F/(ma) =$ idem, or in terms of model and full-scale,

$$F_N/(m_N a_N) = F_M/(m_M a_M). \quad \ldots (3.1)$$

Considering that mass is the product of density and volume, (3.1) may be written as

$$F_N/(\rho_N l_N^3 a_N) = F_M/(\rho_M l_M^3 a_M). \quad \ldots (3.2)$$

If we confine ourselves to solving problems related mainly to weights and stresses generated in rock due to external forces, then acceleration a in the similitude criterion of Newton would have a value equal to acceleration due to gravity g, i.e., $a = g$. Further, by introducing into formula (3.2), notations of specific gravity $\gamma = \rho g$ and stress $\sigma = F/l^2$, implying by it ultimate stresses in materials (ultimate strength under compression, tension, shear and also the deformation characteristics of rock—modulus of elasticity, modulus of shear etc.), equation (3.2) may be written as

$$\sigma/(l\gamma) = \text{idem}; \quad \sigma_N/(l_N\gamma_N) = \sigma_M/(l_M\gamma_M). \quad \ldots (3.3)$$

Three cases arise from criterion (3.3).

1) If the model and the full-scale are made of the same material, then $\gamma_N = \gamma_M, \sigma_M = \sigma_N$ which means, according to (3.3), $l_N = l_M$. Hence the question of modelling does not arise as the requisite condition for modelling is $l_N \neq l_M$ and the realisation of geometric scale $l_c = l_N/l_M$ ought to be fulfilled. In modelling such a case, other additional criteria are considered, for example the effect of external forces, Froude criterion $v^2/lg =$ idem. It follows thus that when $g_N = g_M$ and the geometric scale is arbitrary, $l_c = l_N/l_M$, it is necessary to decrease the rate of occurrence of processes in the model by $l_c^{1/2}$ times, i.e.,

$$v_N^2/(l_N g_N) = v_M/(l_M g_M);$$

$$v_M = v_N\sqrt{l_M/l_N} = v_N/l_c^{1/2};$$

$$v_c = v_N/v_M = l_c^{1/2}. \quad \ldots (3.4)$$

Considering that $v = l/t$, we get the time scale from the preceding equality

$$v_N/v_M = l_c^{1/2} = l_N/l_M t_M/t_N = l_c/t_c; \; t_c = l_c^{1/2}.$$

2) If we consider as before that the model and the full-scale are made of the same material, i.e., $\sigma_N = \sigma_M$ but $l_N \neq l_M$, then for modelling it is

essential that the following condition be fulfilled according to (3.3):

$$\gamma_M = \gamma_N l_N / l_M = \gamma_N l_c. \qquad \ldots (3.5)$$

Consequently, in this case the model should be artificially made heavier by some means or the other. This is realised usually with the help of centrifugal forces, i.e., by rotating in a centrifuge or under linear acceleration of the model in special devices (linear accelerators).

3) If all the quantities embodied in criterion (3.3) are not equal to each other in the model and the full-scale, $\sigma_N \neq \sigma_M, \gamma_N \neq \gamma_M$ and $l_N \neq l_M$, this signifies that a different material was used in the model and selected according to the equality

$$\sigma_M = \sigma_N l_M \gamma_M / (l_N \gamma_N) \doteq \sigma_N / (l_c \gamma_c), \qquad \ldots (3.6)$$

This is referred to as modelling by the method of equivalent materials. Hence for cases (2) and (3) to fulfil Newton's similitude criteria, in physical modelling of phenomena associated with weight, the method of equivalent materials or the method of centrifugal modelling is used. In recent years other methods have also been used, however: methods of photomechanics (photoelasticity, photoplasticity and photocreepage), various modelling methods using rock specimens (optical, X-ray pulse methods etc.). Photomechanic methods are based on relationships derived from the method of equivalent materials since rocks are not dealt with directly but rather optically active materials exhibiting birefringence when subjected to loading. These materials differ from rocks in their mechanical characteristics. Methods of modelling with rock specimens are discussed separately below. Here, we shall detail modelling methods using equivalent materials, centrifugal modelling and photoelasticity as they presently happen to be the best developed amongst the several methods of photomechanics.

3.1.2 Modelling of Stress Waves by the Method of Equivalent Materials

The method of equivalent materials is adopted in investigations related to manifestation of rock pressure, for example, while studying the effect of rock pressure at winning faces and in preparatory workings of mines, in studying the problems of rock swelling, sliding of massifs and other physical processes occurring in massifs. G.I. Kuznetsov was the first to propose this method.

A study of such problems analytically or practically is wrought with considerable difficulties. Hence their methods of modelling have acquired much significance.

The *essence of the modelling method* lies in the fact that using physical models which duplicate the geological structure of a natural deposit, seam or suite of seams through which the workings are driven, physical processes that interest us may be investigated. In such models certain assumptions are made while preserving the geometric conditions of similitude. The results of such investigations are used for evolving practical recommendations for industry.

Newton's mechanical similitude is the basis for the method, which presupposes geometric, kinematic and dynamic similitude, resulting in condition (3.6). It follows that knowing the mechanical properties of the material in the full-scale σ_N and the ratios $l_N/l_M = l_c$ and γ_M/γ_N, the mechanical properties of the model material σ_M may be quantified, which are necessary for satisfying the similitude criteria (3.6), in other words, for choosing the equivalent material. This is the essential feature of the method of equivalent materials.

In selecting equivalent materials usually the set of values of ultimate strength under compression σ_{compr} under tension σ_{rup} or the set of values of cohesion σ_{coh} and angle of internal friction for rocks and soils, is used as the governing characteristic for strength. Here the following equalities are maintained

$$
\begin{aligned}
(\sigma_{compr})_M &= (\sigma_{compr})_N \frac{l_M \gamma_M}{l_N \gamma_N}, \\
(\sigma_{rup})_M &= (\sigma_{rup})_N \frac{l_M \gamma_M}{l_N \gamma_N}, \\
(\sigma_{coh})_M &= (\sigma_{coh})_N \frac{l_M \gamma_M}{l_N \gamma_N}, \\
\tan \varphi_N &= \tan \varphi_M. \qquad \ldots (3.7)
\end{aligned}
$$

The following equalities are to be satisfied for drawing similarities between processes of elastic deformations

$$
\begin{aligned}
E_M &= E_N \frac{l_M \gamma_M}{l_N \gamma_N}, \\
\nu_N &= \nu_M, \qquad \ldots (3.8)
\end{aligned}
$$

where $E_N E_M$ are moduli of elasticity of materials used respectively in the full-scale and the model; ν_N, ν_M are the corresponding Poisson ratio for these materials.

The following condition should be satisfied for the similarity of plastic deformation processes in the entire range of stresses from zero to failure stress without taking into account the time factor

$$
\frac{(\varepsilon_{pl})_M}{(\varepsilon_{el})_M + (\varepsilon_{pl})_M} = \frac{(\varepsilon_{pl})_N}{(\varepsilon_{el})_M + (\varepsilon_{pl})_N}. \qquad \ldots (3.9)
$$

Here, the equality

$$
\varepsilon_{el} + \varepsilon_{pl} = f(\sigma) \qquad \ldots (3.10)
$$

holds good, where $\varepsilon_{el}, \varepsilon_{pl}$ are relative elastic and plastic deformations.

Apart from strength and elastoplastic characteristics for individual layers in the model, similitude of initial and boundary conditions is also ensured. The model should represent a miniature copy of the geological section of rocks in the full-scale from the surface to location of the working (Fig. 9). If it is not possible to represent in the model all the seams present in the full-scale, then at least the seams directly adjoining the working should be represented. The weight of overlying seams is modelled in accordance with condition (3.6) using a special top-loading device, i.e., fulfilling the following condition for the remaining soil layer $(h_{rsl})_N$

$$G_{ld}/S_{ld} = (\gamma_{rsl}h_{rsl})_N/(\gamma_c l_c), \qquad \ldots (3.11)$$

where G_{ld} and S_{ld} are weight of load and area occupied by it; $(\gamma_{rsl})_N$ the specific gravity of the remaining soil layer of the full-scale seam.

A similar rate of advance of the face must also be maintained.

While selecting and determining the scales of modelling, we usually begin with independent arbitrary scales

$$l_c = l_N/l_M; \; g_c' = 1; \; \gamma_c = \gamma_N/\gamma_M.$$

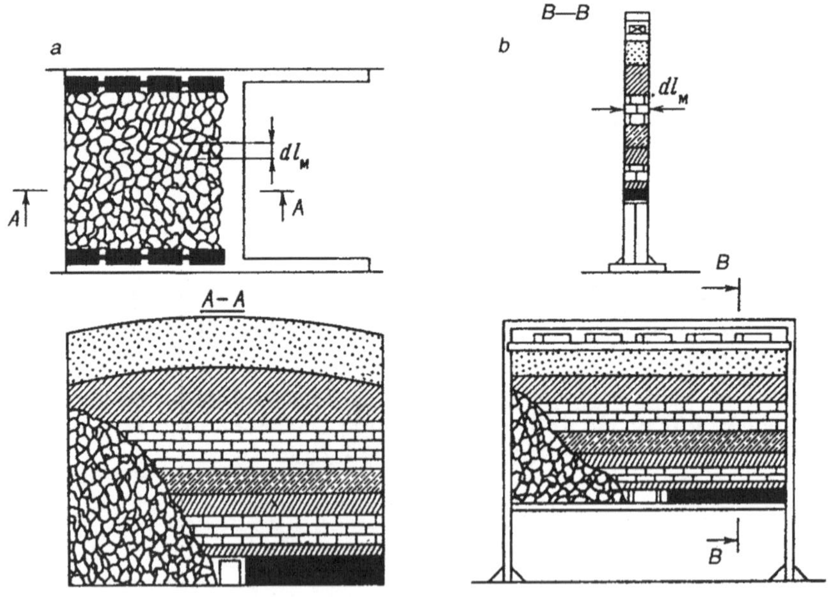

Fig. 9. Scheme for modelling rock deformation by the method of equivalent materials. a—structure of natural seam; b—flat model of such a seam.

The remaining scales are derived from these three arbitrary scales l_c, g_c and γ_c and according to the known constraint equations between them.

Scale of forces

$$F_c = m_c a_c = m_c g_c = m_c;$$
$$g_c = 1;\ m_c = \rho_c l_c^3 = \gamma_c l_c^3 / g_c = \gamma_c l_c^3;\ F_c = \gamma_c l_c^3.$$

Scale of stresses and moduli of elasticity

$$\sigma_c = F_c / l_c^2 = \gamma_c l_c;\ E_c = \sigma_c = \gamma_c l_c.$$

The time-scale for elastic deformation is obtained from the relationship

$$g_c = l_c / t_c^2,\ \text{hence}\ t_c = l_c^{1/2}.$$

In plastic deformations the stress is proportional to the coefficient of viscosity of the material per unit time. In such a case the time scale is given by

$$t_c = t_N / t_M = \eta_N \sigma_M / (\eta_M \sigma_N) = \eta_N l_M \gamma_M / (\eta_M l_N \gamma_N)$$

or $$t_c = \eta_c / (l_c \gamma_c),$$

where η_c is the scale of coefficient of dynamic viscosity.

In the selection of equivalent materials, the conditions of dynamic similitude (3.6) to (3.11) are utilised according to the problem under study. Materials used in the model should conform to certain specifications and physical properties. Such materials are quartzitic sand, fine mica, chalk and clay and binding agents such as gypsum, paraffin, vaseline. The selection process is very scrupulous and requires special equipment. The minimum requirements of apparatus and instruments [9] are as follows: set of vibrating sieves for determining granulometric composition; pneumatic area gauge T-3 for determining specific area of granular materials; vibrating table and manual hammer for preparing test specimens; grinding mill for material grinding; 'Stroiytsnil' cone for determining consistency of material; 'VIK' device for establishing setting time of gypsum; technical and analytical balances; thermostat, thermometers, electric hot plates, set of containers etc.

Highly sensitive instruments are required for determining the mechanical characteristics of materials. For this purpose, specimens taken directly from the layers of the model are used. Pendulum-type force gauges are used to measure forces: for bending, the M.F. Shklyarovsky gauge, for shear, the G.L. Fisenko gauge, and for compression, a 4-tonne laboratory press.

Models are placed on horizontal, rotating stands and also on three-dimensional laboratory fixtures (Fig. 10).

While working on the model, absolute displacements of fixed points in the model and deformations in model elements are measured. Displacements of individual points are recorded by a pair of calibrated microscopes with magnification 20 to 80 times as well as methods of photographic location

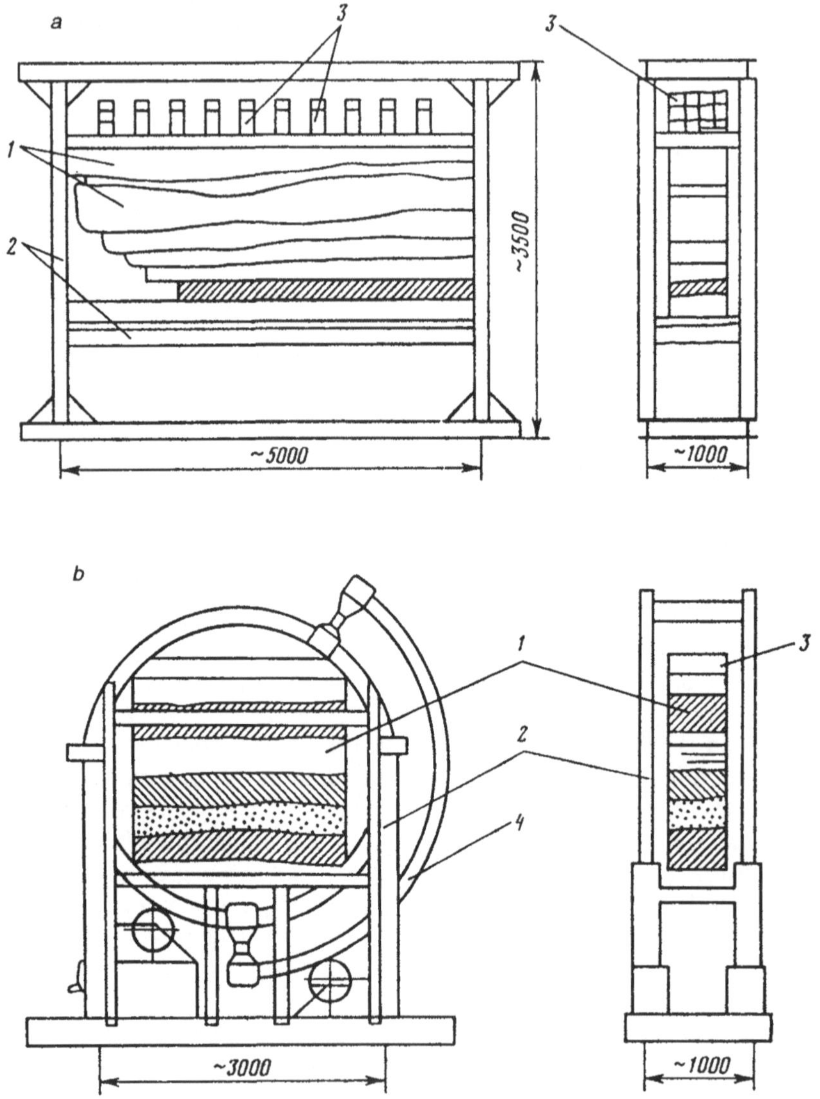

Fig. 10. Laboratory set-up.
a—flat-stationary; b—flat-revolving; 1—model of mineral seams; 2—frame of set-up; 3—additional weight on model; 4—revolving device.

(periodic snaps of surface of the model with specific divisional markings). Deformations in the elements of the model are measured by the two microscopes.

Microdynamometers with electrical (bridge principle) recording are used for determining the stress state of elements in the model. They are initially calibrated.

Attempts have recently been made to use the method of equivalent materials for modelling the effect of a blast on the subsequent behaviour of workings and pillars, their stability and failure and also for modelling the stress waves. In such a case, apart from the requirements of Newton's mechanical similitude (3.1), it is necessary to take into account conditions of dynamic similitude of blast effect, leading to selection of such an explosive as will induce an effect similar to that of a blast in the full-scale conditions.

Let us consider the peculiarities in modelling by this method, the slabbing phenomena and damage to the walls of the working due to the effect of blast-induced stress waves, and also the modelling of kinematics of rock particle motion induced by the piston effect of detonation products. The first two phenomena are determined not only by the maximum value of stresses, but also by the duration and shape of a stress-wave diagram. The thickness of a slab/chip may be expressed roughly by the following relationship

$$\delta = \frac{\sigma_{rup}}{2\sigma_{max}} c_p(\tau_+ - \tau_N),$$

where σ_{rup} indicates the tensile strength of rock; σ_{max} the maximum amplitude of stress wave; τ_+, τ_N duration of positive phase and duration of growth of stress wave amplitude to the maximum respectively. Hence under geometric similitude $\delta_N = l_c\delta_M$, the following relationship [5, 7] may be obtained

$$\left(\frac{\sigma_{rup}}{\sigma_{max}}\right)_N = (c_p\tau_+)_N\left[1 - \left(\frac{\tau_N}{\tau_+}\right)_N\right]$$

$$= l_c(c_p\tau_+)_M\left(\frac{\sigma_{rup}}{\sigma_{max}}\right)_M\left[1 - \left(\frac{\tau_N}{\tau_+}\right)_M\right].$$

From this the following requirements are prescribed for modelling parameters that determine the maximum amplitude, shape and duration of stress wave,

$$\left.\begin{aligned}&\left(\frac{\sigma_{rup}}{\sigma_{max}}\right)_N = \left(\frac{\sigma_{rup}}{\sigma_{max}}\right)_M, \left(\frac{\tau_N}{\tau_+}\right)_N = \left(\frac{\tau_N}{\tau_+}\right)_M,\\ &(c_p\tau_+)_N = l_c(c_p\tau_+)_M \quad \text{or} \quad \lambda_N = l_c\lambda_M.\end{aligned}\right\} \qquad \ldots (3.12)$$

Considering the conditions for modelling by the method of equivalent materials (3.6) and $g_c = l_c/t_c^2 = 1$, it is easy to obtain the scales for modelling the indicated parameters of stress wave from the requirements of (3.12):

$$(\sigma_{max})_N/(\sigma_{max})_M = (\sigma_{rup})_N/(\sigma_{rup})_M = \sigma_c = \gamma_c l_c,$$

$$(\tau_N)_N/(\tau_N)_M = (\tau_+)_N/(\tau_+)_M = t_c = l_c^{1/2},$$

$$(c_p)_N/(c_p)_M = l_c(\tau_+)_M/(\tau_+)_N = l_c/t_c = l_c^{1/2}. \quad \ldots (3.13)$$

It follows from these relationships that the scale of amplitude of stress wave should be equal to the scale of strength characteristics of rocks both in the full-scale and the model, while the time scale should be lower in the model when charges are blasted in the same rock where $t_c = l_c$ and the scale of propagation rate of the wave should be equal to the time scale. Here the scale of rate of displacement in the stress wave is obtained from $\sigma_m = \rho_p\ c_P v_m$; expanding it, we have $(\rho_p\ c_P v_m)_N/(\rho_p\ c_P v_m)_M = (\gamma c_P v_m)_N/(\gamma c_P v_m)_M = \sigma_c = \gamma_c l_c$, and considering from (3.13) that the scale for $(c_P)_c = l_c^{1/2}$, we have

$$\gamma_c l_c^{1/2} v_c = \gamma_c l_c; \quad v_c = l_c^{1/2}. \quad \ldots (3.14)$$

It follows from (3.13) and (3.14) that while modelling by the method of equivalent materials the parameters of stress waves in the phenomena of slabbing and caving of workings, the extended law of geometric similitude valid for modelling the stress wave in one and the same medium does not hold good. The solution lies in choosing such an *EC* that, when blasted will induce waves having parameters satisfying conditions (3.13) and (3.14), i.e., apart from maintaining the geometric similitude of the dimensions of working, geological cross-section of seams and charge columns, it is necessary to select such a source of blast, generating waves of low velocity, while the stress wave itself is of low intensity and longer duration compared to that emitted in a blast of geometrically reduced charge under field conditions. This, obviously, can be achieved by using either an *EC* with less shattering effect or charges of special configuration with an annular gap filled with air or some porous materials and also by using new *EC* in combination with different materials that address the conditions of modelling. In such cases the final solution to the problem requires additional theoretical and experimental investigations.

While modelling the patterns of stress-wave propagation, as shown in Sec. 2 and in [3, 4], the similitude criteria of Euler, Cauchy and Strouhal number should be equal

$$\sigma/(\rho v^2) = \ni, \ v/\sqrt{E\varepsilon/\rho} = \mathrm{Ca}, \ vt/l = H.$$

From the criteria of Euler and Newton, we get the Froude criterion after dividing the former by the latter,

$$[\sigma/(\rho v^2)/\sigma/(\gamma l)] = \gamma l/(\rho v^2) = gl/(v^2) \text{ or } v^2/(gl) = Fr.$$

The velocity scale is readily obtained from the above after expanding it through corresponding quantities for the full-scale and the model:

$$v_N^2/(g_N l_M) = v_M^2/(g_M l_M) \rightarrow v_N/v_M = v_c$$
$$= \sqrt{l_N g_N/(l_M g_M)} = \sqrt{l_c g_c} = l_c^{1/2},$$

which coincides with formula (3.14) and does not contradict the method of equivalent materials for elastic deformations when $v_c = l_c/t_c = l_c^{1/2}$ is derived from the conditions $g_c = 1, t_c = l_c^{1/2}$.

By multiplying the criterion of Cauchy with that of Newton (both criteria being applicable to the modulus of elasticity), the following criterion may be obtained

$$v^2 \rho E/(\varepsilon E \gamma l) = v^2 \rho/(\varepsilon l \gamma) = v^2/(\varepsilon l g) = \text{idem}.$$

Dividing the criterion obtained above by the Froude criterion, we get

$$v^2 lg/(\varepsilon lg v^2) = 1/\varepsilon = \text{idem, or } \varepsilon_N = \varepsilon_M. \qquad \ldots (3.15)$$

This means that the relative deformations of materials in the model and the full-scale should be equal. Condition (3.15) is in conformity with the condition related to the method of equivalent materials.

Finally, from the equality of the Strouhal number and considering the velocity scale $v_c = l_c^{1/2}$, it is easy to obtain the time scale, analogous to (3.13):

$$vt/l = \text{idem}; \; v_c t_c/l_c = 1; \; t_c = l_c/v_c = l_c^{1/2}.$$

Thus in modelling patterns of stress-wave propagation by the method of equivalent materials, the solution also involves selection of an *EC* or charge configuration in accordance with the above-noted features.

In conclusion, let us briefly deal with equivalent material modelling of zones of breakage and crushing. This task is very complex as it involves modelling structural heterogeneities and jointing which is difficult in practice to replicate, due to limited technical possibilities as well as inadequate knowledge of these characteristics in the full-scale. That is why modelling of these parameters of blast-induced effects is always approximate and is influenced by a specific scale effect.

It should be noted that the crushing process is determined by the effect of stress waves and quasi-static pressure of explosion products, creating a grid of new cracks and opening existing microcracks in the first stage of a blast, the preliminary breakage of rocks. It is further determined by the piston effect of detonation products on rock in the second and third stages of a blast, when fragmented rock is pushed towards the free face forming a muck pile. Equivalent material modelling of the first stage of explosion has already been discussed.

Let us consider the modelling of second and third stages of a blast by this method—displacement of rock mass by the piston effect of detonation products and formation of muck pile or cone of throw. In this case, apart from considering the similitude criteria of Euler, Cauchy and Strouhal number, it is essential that the Froude criterion be satisfied which, as shown above, is evolved from the criteria of Euler and Newton while modelling stress waves by the method of equivalent materials. This exercise leads us to the aforesaid conditions for stress waves, i.e., to the task of selection of *EC*, charge configuration or any other source of disturbance in place of a blast.

Thus while modelling the propagation and action of stress waves and piston effect of detonation products by the method of equivalent materials, the law of extended geometric (energetic) similitude is violated, which is valid for a model fabricated from materials of the full-scale. In both cases investigations should be conducted for selection of the source of disturbance that simulates the effect of a blast in the full-scale. In doing so, conditions (3.13) and (3.14) should be satisfied not only for the stress waves, but also for residual pressure in the blast cavity p_0 and duration of its effect τ_p. In the latter case the relationships are of the following type

$$(p_0)_M = (p_0)_N \frac{1}{\gamma_c l_c} = \sigma_c (p_0)_N;$$

$$(\tau_p)_M = \frac{(\tau_p)_N}{l_c^{1/2}}.$$

When l_c and γ_c are arbitrarily chosen (the latter tries to attain a value as near to unity as possible, $\gamma_c = 1$), the parameters of blast source should satisfy the following conditions:

$$(p_0)_c = (\sigma_{max})_c = \gamma_c l_c \approx l_c; \; v_c = l_c^{1/2},$$

$$(\tau_p)_c = (\tau_N)_c = (\tau_+)_c = t_c = l_c^{1/2},$$

$$(I_+)_c = \sigma_c t_c = \gamma_c l_c^{3/2} \approx l_c^{3/2};$$

$$(E_+)_c = \gamma_c l_c^2 \approx l_c^2. \qquad \ldots (3.16)$$

It follows, therefore, that the amplitude of stress wave should be at least less than that in the full-scale to the extent of the scale of geometric similitude, while the particle velocity and all time-related characteristics should be less than those in the full-scale by a quantity equal to the square root of geometric scale $l_c^{1/2}$, the flux of wave energy should be equal to the square of geometric similitude while the stress wave pulse should be lesser by $l_c^{3/2}$ times. When *EC* of the type TEN, Hexagene, Trotyl etc. are used, such conditions are difficult to fulfil, as shown experimentally. In this case the magnitudes of velocity for various rocks in the middle- and far-field zones of a blast are approximately equal and obey the same law of propagation over distance. This confirms that $v_c \approx 1$ and $v_c \neq l_c^{1/2}$, as shown in (3.16). Consequently, special investigations are needed to identify new types of *EC* or charge configurations having an annular gap filled with a special loose substance and which will reduce the intensity and extend the pulse of stress wave in the equivalent material.

3.1.3 Centrifugal Modelling of Blast Effects in Rocks

The centrifugal modelling method, like the method of equivalent material, is mainly used for studying problems related to rock pressure (investigations into slope stability, stresses in soils underlying foundations, settlement of structures etc.). Currently, it is also being used for the study of blast-induced stress fields, the phenomena of muck displacement and fly rock.

The method involves preparation of a model from the same material as found in the full-scale, in case geometric similitude is satisfied; placing the model in a centrifuge; and generation of a high rate of acceleration in the model under the effect of centrifugal forces to provide the mechanical similarity of forces acting on the full-scale.

Usually the model dimensions are determined by the dimensions of the centrifuge and hence this method has practical constraints, in particular, working with small models leads to difficulty in maintaining the geometric similitude of rock structure, its lamination etc., while in working with large models difficulties arise in fabrication of large centrifuges. Given such problems the method of centrifugal modelling is rarely used. Sometimes, it is adopted in combination with the method of equivalent materials.

The centrifugal modelling method, like the method of equivalent material, is based on the theory of Newton's mechanical similitude and results in condition (3.5) being satisfied, i.e., leads to the condition wherein for maintaining the condition of similarity the specific gravity of material of the full-scale used in the model should exceed that of the full-scale by the same number of times the size of the full-scale exceeds that of the model. To achieve this the value of γ_M is enhanced by inertial forces, without effecting any change in the physicomechanical properties of material, by rotating the model at a certain acceleration a. Then we have $\gamma_M = \rho a; \gamma_N = \rho g$. From (3.5) we obtain

$$\gamma_M = \gamma_N l_c, \quad a = l_c g, \qquad \ldots (3.17)$$

i.e., the model should be subjected to an acceleration above that of gravity by l_c times (by the magnitude of the linear scale).

The scalar quantity of total acceleration of the model placed in a centrifuge having a vertical axis is given by

$$a = \sqrt{g^2 + (\omega^2 R)^2}, \qquad \ldots (3.18)$$

where g is acceleration due to gravity; ω angular velocity; R distance from a given point in the model to the rotational axis of the centrifuge.

It is easy to obtain the values of ω and n from (3.17) and (3.15):

$$\omega = \sqrt{g/R}\,\sqrt[4]{l_c^2 - 1},$$

$$n = \frac{30}{\pi}\sqrt{g/n}\,\sqrt[4]{l_c^2 - 1}, \qquad \ldots (3.19)$$

where n is the number of revolutions per minute, equal to $\frac{60}{2\pi}\omega$.

From formula (3.18) it is clear that each point of the model, situated on its own radius vector R, should rotate with individual angular velocity ω, which cannot be practically realised in a centrifuge. In this case the error would be determined by the variation in acceleration a when R is changed. From formula (3.18) we have

$$a_1 = \sqrt{g^2 + \omega^4 (R+h)^2},\, a = \sqrt{g^2 + \omega^4 R^2},$$

$$\Delta a/a = (a_1 - a)/a = a_1/a - 1$$

$$= \left(\sqrt{g^2 + \omega^4 (R+h)^2}\sqrt{g^2 + \omega^4 R^2}\right) - 1;$$

as $\omega^2 R \gg g$, the error is approximately computed as

$$\delta a = \Delta a/a \approx \frac{\omega^2 (R+h)}{\omega^2 R} - 1 = \frac{R+h}{R} - 1 = \frac{h}{R}. \qquad \ldots (3.20)$$

Thus centrifugal modelling is only an approximate modelling method. Here, the smaller the model dimensions vis-à-vis the radius of rotation, the smaller the error.

Usage of equivalent materials in the method of centrifugal modelling may reduce the angular velocity of rotation of the centrifuge ω when $\gamma_M > \gamma_N$, which means that by increasing the dimensions the error δa may be reduced. The condition $\gamma_M < \gamma_N$ leads to an increase in l_c by $\gamma_N/\gamma_M > 1$, which is associated with an increase in ω and reduced dimensions of the centrifuge. The latter measure again results in an increase in error δa. In both these cases the use of equivalent materials is not advisable for investigating blast-induced effects, as such effects are related to the problem of selection of *EC* or charge configuration (which has not been studied). Thus additional errors and technical difficulties would result. It is therefore advisable to use rock specimens scaled down by the magnitude of geometric similitude l_c as the model while modelling a blast.

The actual radius of rotation of the centrifuge R_{cen} should be equal to 2–3 m, according to practical considerations and accumulated experience of using this method for studying blast effects.

According to G.I. Pokrovsky, due to the assumptions made (neglecting the dimension of model in comparison to R_{cen}), the errors of observation would increase proportional to the scale of geometric similitude, i.e.,

$$\delta_{obs} = l_c \delta_{mse}$$

where δ_{obs} is the mean square error in observations; l_c the geometric scale of similitude; δ_{mse} the mean square error in the adopted method of measurement.

From the assessments of accuracy, as conducted above, it follows that centrifugal modelling is good by and large for static problems, such as distribution of stresses and deformations, where it is possible to reduce considerably the errors in measurement δ_{mse}. For dynamic deformations and caving in the model, this method may yield large distortions and require further special investigation and substantiation. Centrifugal modelling gives better results in the modelling of problems related to filtration and viscous deformations, as the velocity of the process in the model enhances by l_c times and the time itself by $l_c^{1/2}$, i.e., there is a perceptible rapid stabilisation of wet soils and viscous processes.

Currently, centrifugal rockers, revolving around a vertical axis, are used. In future, more modern centrifuges of another type might be used, for example a circular one in which the model is placed in a chute moving in an annular chamber as on a closed circular rail track.

It is to be noted that similar to the method of equivalent material modelling, loads are used in centrifugal modelling also, if all the seams above the working cannot be duplicated on the rotor of the centrifuge or if the height of the model becomes too large. By this means the height of the model is reduced, which favourably affects the error in modelling. Finally, the friction between rock and wall of carrier of centrifuge also affects errors in modelling. To reduce such errors the internal walls of the carrier are greased, while the carrier itself is made in the form of a truncated cone so that when rocks are whirled, a gap is formed, due to which lateral pressure is avoided which might otherwise distort the stress state of the rock.

As of today, two groups of modern centrifuges are distinguished: large (Fig. 11, a) and small (Fig. 11, b). The former are installed in special basements and the radius of suspension of the model is 2–3 m from the rotational axis. Small centrifuges are placed in enclosures and the suspension radius of the model is 1 m from the rotational axis.

The modelling of a foundation settlement in soil is shown schematically in Fig. 12.

Modern centrifuges are equipped with special slip rings enabling transmission of readings of several gauges of various systems intended for various purposes and also control of the load variation on the model.

Centrifuges are equipped with controlling and measuring instrumentation, linked to the rotational axis of the centrifuge, and also with devices and instruments used in the modelling of static and dynamic processes in rock.

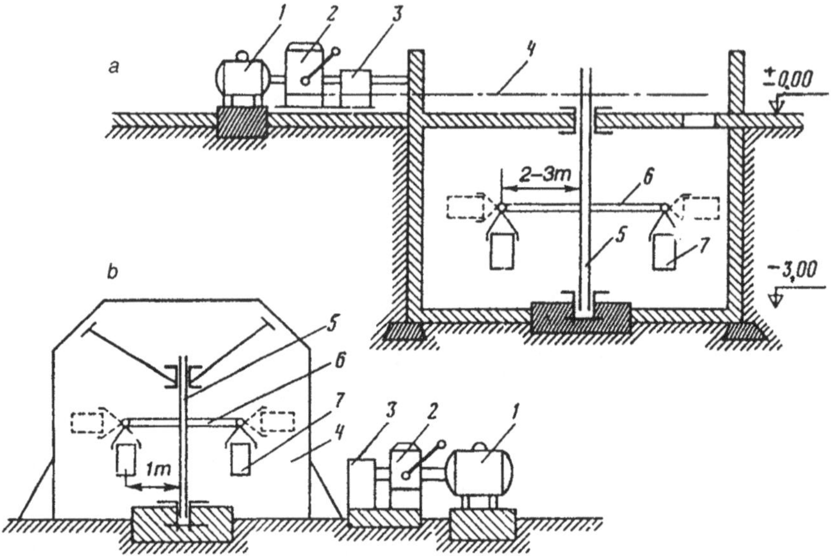

Fig. 11. Constructional features of a centrifuge.
1—motor; 2—speed controller; 3—reduction gear; 4—transmission of rotation from motor to axis (5); 6—rocker; 7—baskets for placing models.

The simplest device for measuring the settlement of soil is the sliding scale, consisting of a graded metallic rod moving in a special casing with the help of a micrometer. It is used for measuring the profile, volume and slope of the model. A profiler is also used for such purposes (two sheets, round metallic rods and clamps). Tentative deformations of profile may be determined by using it, with an accuracy of about 0.2 mm.

Various dynamometers are used for measuring stresses in a soil model, soil pressure on the resting model wall or mine support. For example, the aerostatic dynamometer consists of an oval-shaped box filled with coloured glycerin, a capillary and a diaphragm. Pressure is computed applying Claperion's law. Dynamometers with various strain gauges, even foil type strain gauges are used.

Synchro-devices are used for measuring deformations and displacements—self-synchronising electrical micromachines for smooth synchronisation of the transmission of rotational angle to the shafting of another micromachine some distance away. The extent of settling is measured by a remote automatic recorder.

Ten electric pulse counters, energised by pulses of direct current through circuit switching in an electrical circuit breaker, are used for recording the breakages that occur in the model and the time of their initiation. The extent

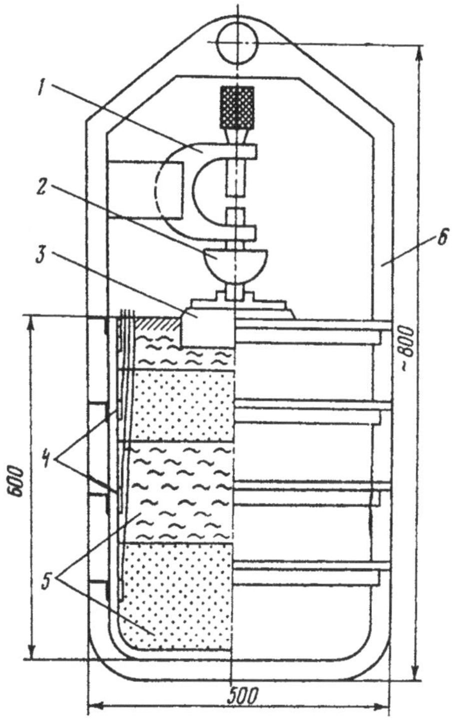

Fig. 12. Centrifugal modelling of foundation settlements in soils.
1—micrometer; 2—mercury container; 3—press; 4—dynamometers; 5—models of soil layers; 6—baskets of centrifuge.

of breakage is gauged by 2-mm wide strips of a perforated tin sheet placed at the expected location of breakage.

For measuring deformations in models, induction devices of various configurations and strain gauges of low weight base (2–4 mm) are also used.

All the instruments may be divided into two major groups: devices placed in rock, displacing a part of its volume, belong to the first group, while devices measuring deformations in certain elements of the rock within the massif belong to the second group.

Stroboscopes in combination with cinematography and television are used for visual observations. A stroboscope is a relatively simple device capable of fixing fast-occurring periodic processes with the help of a reading magnifier. Vibrations and deformations in the model can be monitored by means of a stroboscope.

In centrifugal modelling of phenomena induced by a blast in rocks, fulfilment of the similitude criterion of Euler and the Strouhal number for similar media in the model and the full-scale leads to the extended law of geometric similitude for stress waves as formulated in [5]. Here the Cauchy criterion is automatically fulfilled, while the Froude criterion—necessary for phenomenon related to the piston effect of gases, i.e., phenomenon of formation of a cavity, scatter, flying and muck piling of a rock mass—is fulfilled by acceleration (3.18) generated by rotation of the centrifuge.

Consequently, modelling in this case leads to selection of the scale of geometric similitude, wherein the technical capabilities of the centrifuge are to be kept in view: its radius, range of variation of angular velocity and maximum charge weight in the model at which stable detonation is still feasible.

The capabilities of a centrifuge may be assessed by formula (3.19), from which a relationship for the maximum scale of geometric similitude is readily obtained:

$$l_c \leqslant l_{c_{max}} = \left(\frac{\omega_{max}^4 R_{max}^2}{g^2} + 1\right)^{1/2} \approx \frac{\omega_{max}^2 R_{max}}{g} = \frac{\pi^2 n_{max}^2 R_{max}}{900g},$$

where ω_{max} or n_{max} is the maximum angular velocity or number of revolutions of the centrifuge; R_{max} the maximum distance from the model to the axis of rotation.

From this relationship we can derive the required angular velocity for the chosen l_c or given R_{max} and also the maximum radius of centrifuge $R_{cen} = R_{max} - \Delta R$ under known values of l_c, ω_{max} and ΔR, distance from end of rocker to the point being studied in the model

$$R_{cen} = (g/\omega^2)\sqrt{l_c^2 - 1} - \Delta R \simeq g l_c/\omega^2 - \Delta R.$$

3.1.4 Photoelastic Method

The photoelastic or optical-polarisation method is mainly used for studying the stress state in structures, machines, buildings and machine parts. More recently it has been used for studying the stress distribution around mine workings, investigating rock breakage under the action of a drilling tool and blasting.

To study the stress state in rocks, a model is prepared from optically active material by maintaining geometric energetic similitudes. Polarised light rays are transmitted through this model and the stress field revealed on a screen in the form of lines of various colours and shades. Both two-dimensional (more so) and three-dimensional models are currently used. The latter are preferred for studying rock breakage under the action of a drilling tool and blasting.

Theoretical principles of the photoelastic method are based on Hooke's law regarding the direct proportionality between stress and strain and also the phenomenon of optical anisotropy of transparent isotropic bodies under the action of external forces or due to internal stresses present in them. Here the values of normal stresses σ_1 and σ_2 are not of interest, but the difference between them or the maximum shear stresses is determined,

$$\tau_{max} = (\sigma_1 - \sigma_2)/2, \qquad \ldots (3.21)$$

which is a key characteristic of structural strength according to Coulomb's theory.

The general theory of light and theory of light polarisation are the basic premises of the optical method. In polarised light the light vector vibrates in a particular plane (plane polarised light) or in an elliptical manner (elliptically polarised light), and does not vibrate along irregular directions in the plane perpendicular to the ray of light, which usually happens with ordinary light. When an ordinary light ray falls on a glass at an angle, the resultant reflected and refracted rays would be plane polarised and would vibrate in two mutually perpendicular planes (Fig. 13). In this case, if the reflected and refracted rays are perpendicular to each other, maximum polarisation is achieved. Some crystals (tourmaline, mica, calcite) possess the property of double-refraction (ordinary and non-white rays). Both rays happen to be plane polarised; vibrations in the ordinary ray are pe"endicular to the diagonal plane *aa'cc'*, while those in the non-white light lie in the diagonal plane (Fig. 14). Such a phenomenon is also observed in isotropic substances subjected to loading. This particular aspect is used in the photoelastic method. The property is exhibited by glass, resir plastic, celluloid etc., when subjected to loading, and disappears on removal of the load.

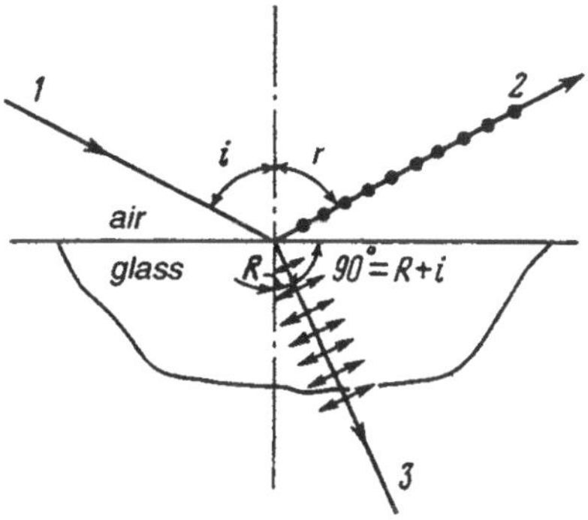

Fig. 13. Reflection and refraction of light rays of boundary of air-glass.
i, *r* and *R*—angles of incidence, reflection and refraction respectively; 1, 2, 3—incident, reflected and refracted rays respectively.

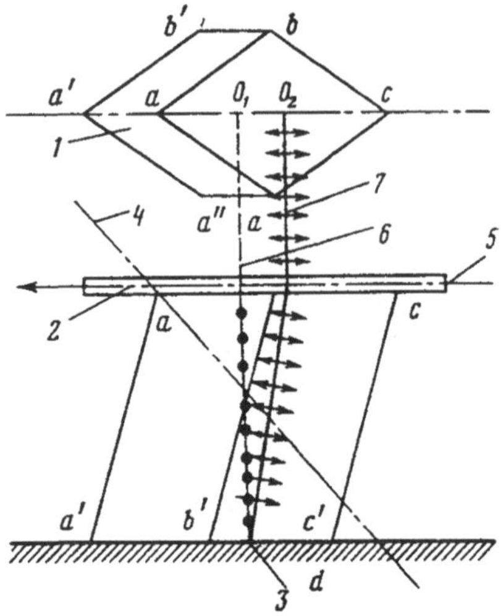

Fig. 14. Double refraction through a calcite crystal.
1—calcite crystal; 2—tourmaline sheet; 3—marker; 4—optical axis of calcite crystal; 5—polarisation axis of tourmaline sheet; 6—white (ordinary) ray; 7—non-white (extraordinary) ray.

It is known that in a two-dimensional stressed state there exists an area at each point in which shear stresses are absent and only the principal normal stresses σ_1 and σ_2 act. Under the action of the latter, corresponding deformations occur that lead to optical anisotropy in the material. A ray passing through such material splits into two plane polarised rays, in which vibrations occur in mutually perpendicular planes. The two polarised rays themselves run parallel to the direction of propagation of the incident ray and pass through the corresponding normal stressed zones. These rays progress with differing velocities based on σ_1 and σ_2, which results in optical shifting of phases and a difference in linear path (G). This is determined according to the following formula [9] based on a large number of experiments:

$$G = C_\sigma \delta (\sigma_1 - \sigma_2), \qquad \ldots (3.22)$$

where C_σ is the stress-optic coefficient; δ the sheet thickness; σ_1, σ_2 principal normal stresses.

Considering formula (3.22), expression (3.21) may be rewritten as:

$$G = 2C_\sigma \delta \tau_{max}. \qquad \ldots (3.23)$$

The relative difference G is determined on the basis of the property of polarised light to exhibit different fringes depending on frequency of vibrations (just as the pitch of sound is determined by frequency of sound vibrations).

It is known that for each monochromatic light, there exists a specific wavelength

$$\lambda = vT = v/p_K; \ \lambda = c^*/p_K,$$

where v and c^* are velocity of light propagation through material and vacuum; T and p_K the time period and frequency of vibration of wave. Thus, after obtaining different colours on the screen and knowing in advance the relationship between colour and wavelength, the relative difference $G = n\lambda$ (where n is the sequence fringe number) and τ_{max} from (3.23) are readily determined, i.e.,

$$\tau_{max} = n\lambda/(2C_\sigma\delta) = \tau_0^{1.0}n/\delta, \qquad \ldots (3.24)$$

where $\tau_0^{1.0}$ is the optical constant of the material (rating of the band), which is the magnitude of stress required for varying the sequence of band n by one unit and is determined in advance on the specimen of model material by the method of comparison of colours of bands and method of compensation.

In the method of fringe width comparison, the unknown relative difference is determined directly from the isochrome colour chart obtained from an ordinary (white) source of light by comparing the colours with the interference fringes of standard Newton rings, for example, based on the Table given in [9], where the band sequence, serial number of fringe with its characteristic and relatively differing rays (in nanometers) are given. The method of bands is adopted when a monochromatic light source is used. Here the fringe number in the isochrome and rating of the relevant strip of material are established, based on those obtained in advance while calibrating the selected optically sensitive materials.

The modulus of elasticity E and Poisson ratio ν are determined by measuring deformations in specimens of the material using a mechanical strain gauge. Specimens are in the form of extended strips. Load is applied in stages and computations are done by the formulae

$$E = mS\sigma/\varepsilon_y; \ \nu = -\varepsilon_x/\varepsilon_y,$$

where m and S are scale and nominal length of the strain gauge; σ the stress in specimens; $\varepsilon_x, \varepsilon_y$ the relative deformations in the direction of x and y axes.

Rating of the strip of material is established according to its sequence at the centre of a 1-cm thick disc, compressed diametrically by a directed force P, by the formula

$$C_\sigma^{1.0} = 8P/(\pi dn_c) \approx 2.548P/(dn_c),$$

where n_c is the actual number of the strip at the centre of the disc; d the diameter of the disc.

The quality of model material is characterised by the degree of deformation of the model for obtaining the required optical effect and is assessed by the quality index

$$K = (E/C_{\sigma}^{1.0})10^{-3}.$$

The higher the value of K, the lower the deformation needed in the specimen to induce the same optical effect.

Calibration is done on a special calibrating press (Fig. 15). Tentative calibration curves are given in Fig. 16.

The calibrating press enables application of a uniform load on the model material (3) by pouring water into bowl (1) and corresponding by balancing (1) by pouring water into bowl (5). Bowls (1) and (5) are suspended at the edges of a balancing beam (2), which itself is pivoted to column (4). The

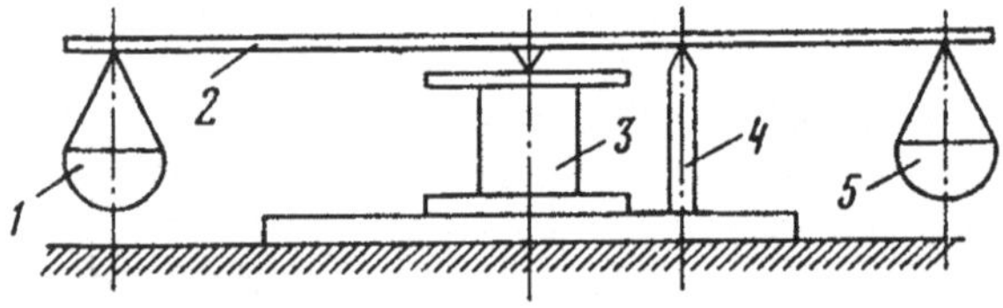

Fig. 15. Calibration press for optical materials.

1 and 5—bowls; 2—balance be..n; 3—model specimen; 4—pivoted column.

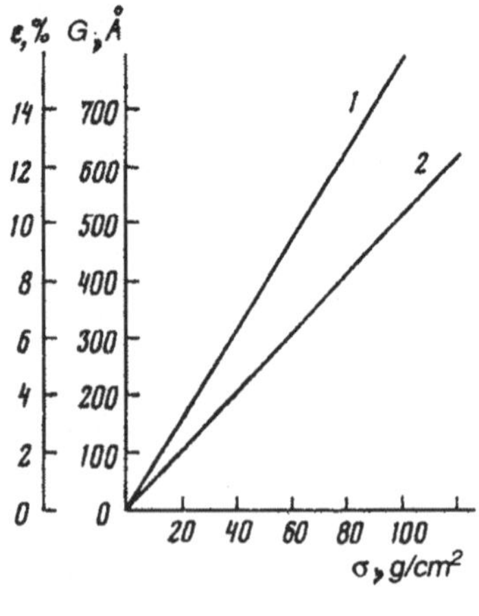

Fig. 16. Calibration curves.

1—relative difference $G = f_I(\sigma)$; 2—deformation $\varepsilon = f_2(\sigma)$.

stress-optic coefficient is computed from the data of relative difference of rays G and the measured stresses σ_1 (deformation $\varepsilon = E/\sigma_1$),

$$C_\sigma = G/(\sigma_1\delta)$$

or the optical constant $\tau_0^{1.0}$ is calculated using formula (3.24) based on the recorded number of n bands passing through a monochromatic light source.

In the method of compensation, a known relative difference with opposite sign is added by the compensating device to the obtained relative difference in the loaded model specimen, such that when crossed polaroids are installed, the point of measurement in the specimen remains dark, which confirms the equivalence of relative retardation in the loaded specimen and that of the compensator.

Having dealt with the preliminary concepts, let us now consider the principle of operation of a polariscope (Fig. 17). Based on this Figure, the resultant amplitude of vibrations after the analyser is given by

$$A_z = a_1 + a_2 = A_0 \sin\alpha \cos\alpha[\sin(\varphi - 2\pi n) - \sin\varphi], \qquad \ldots (3.25)$$

where A_0 is the amplitude of vibrations of a single ray after passage through the polariser; a and a_2 are amplitudes of vibrations of rays after passage through the analyser; α is the angle between the axis of specimen and the horizontal axis; φ is the angle of phase shift of white and non-white rays, formed after the polarised ray with an amplitude of A_0 passed through the loaded model 3.

At

$$\varphi = -\pi n$$
$$A_{max} = A_0 \sin(2\alpha) \sin(\pi n).$$

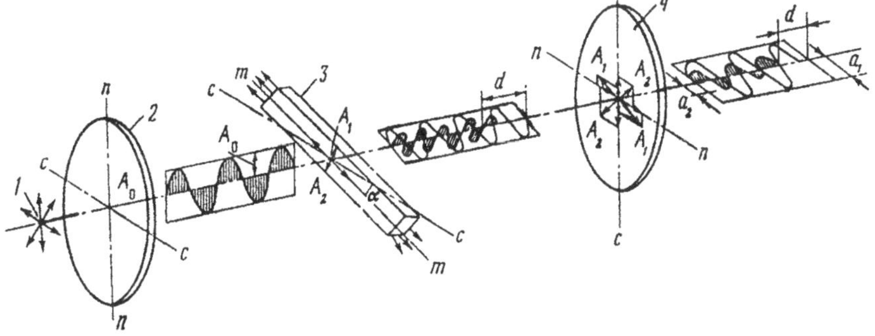

Fig. 17. Schematic diagram of a polariscope.
1—light source; 2—polariser; 3—specimen (model); 4—analyser; *c-c*—horizontal axis of polariser; *n-n*—vertical axis of polariser; *m-m*—axis of model; *d*—path difference in rays; A_1 and A_2—amplitudes of polarised rays.

Light intensity is given by

$$I = A_{max}^2 = A_0^2 \sin^2(2\alpha)\sin^2(\pi n).$$

Maximum intensity is observed at $\alpha = 45°$. Here the colours are seen distinctly (or bright fringes in the case of monochromatic light, isochromes). Apart from the bright colours, dark fringes and dark spots appear on the screen, satisfying one of the three conditions given below.

1. Locations, free from stresses (the analyser blanks the polariser): this usually indicates the entire field before a blast.

2. Points where $\sigma_1 = \sigma_2$, i.e., there is no phase shift. This means, from formula (3.25), that

$$A_z = a_1 + a_2 = A_0 \sin\alpha\cos\alpha\sin\alpha + (-A_0 \sin\alpha\cos\alpha\sin\alpha) = 0.$$

3. Points for which the directional planes of principal stresses coincide with the planes of polariser and analyser, i.e., at $\alpha = 0$ or $\alpha = 90°$, we have $I = 0$, as $\sin(2\alpha) = 0$.

The third case is of paramount importance and is used for determining the direction of two principal stresses. In this case either the specimen is rotated or the system analyser-polariser is rotated, and the isoclines—system of dark spots combining to form a dark band—are located. These vary with varying angle of rotation of the analyser (the polariser). Thus, for each angle of rotation of the system analyser-polariser there is a corresponding specific dark line, an isocline. Based on it isostats are plotted against it. Isostats are directional curves of one of the principal stresses σ_1, σ_2. They are mutually perpendicular, i.e., they represent an orthogonal family of curves whose tangents coincide with the direction of principal normal stresses. The principle of plotting isostats is shown in Fig. 18.

Isoclines 1, 2, 3 are shown in the figure, which correspond to angles of rotation α_1; $\alpha_2 = \alpha_1 + \Delta\alpha$ and $\alpha_3 = \alpha_1 + 2\Delta\alpha$. From points *A*, *B* and *C* of these isoclines, lines at angles α_1, α_2 and α_3 to the vertical are drawn which intersect at points *D* and *E*. The broken line *ADBEC* thus obtained is the isostat-directional line of one of the principal stresses, for example σ_1. At each point in the isostat, the principal stress σ_2 is directed at 90° to the isocline σ_1. Thus, the isostats constitute a family of curves whose tangents coincide with the direction of principal normal stresses. Here it should be remembered that the isoclines are generated by the dark spots which vary with rotation of the analyser-polariser system, as distinct from the unstressed places or points at which $\sigma_1 = \sigma_2$, which do not change with rotation of the analyser-polariser system.

In this manner the following results are obtained from a study of a specimen or model in a polariscope: the value of τ_{max} in the zone of elastic deformations (isochromes, the bright, coloured bands), fields of stresses σ_1 and σ_2 and, consequently, the zones of low and high stresses. After fabrication, the flat model is installed in the path of polarised light rays and the field

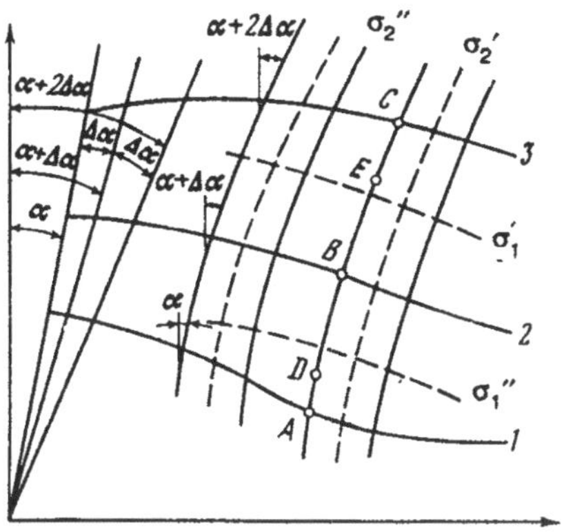

Fig. 18. Plotting isostats.
$\sigma_1', \sigma_2', \sigma_1'', \sigma_2''$—isostats.

of distribution of colours corresponding to the tangential (shear) stresses is observed. With an increment in load, the picture of distribution of colours on the screen changes up to a certain specific value depending on the angle of phase shift; the picture is later repeated in the same sequence several times and the coloured bands of first, second and third orders etc. recorded. The sequence of these bands (number of bands) is directly proportional to the increase in tangential (shear) stress in accordance with formula (3.24).

The nature of distribution of colours and their variation under loading may be photographed by separate films or videotaped or plotted indicating the sequence of bands.

To plot the field of principal normal stresses, the field of isoclines is initially drawn by rotating the polariser-analyser frame from $\alpha = 0$ to $\alpha = 90°$ with reference to the axes of the model or specimen (for solving static problems). This field is recorded by the same methods as the field of isopanes (lines of equal tangential stresses) or isochromes (lines of principal stresses).

To determine the principal stresses, we should also know their ratio, i.e., σ_1/σ_2.

This ratio in the first approximation for elastic waves may be expressed as

$$\sigma_1/\sigma_2 = 1 - 2\gamma^2 = \nu/(1-\nu), \qquad \ldots (3.26)$$

where $\gamma = v_S/v_P = \sqrt{\dfrac{1-2\nu}{2(1-\nu)}}$; v_S is velocity of propagation of transverse waves; v_P velocity of propagation of longitudinal waves; ν Poisson ratio.

The foregoing discussion concerns modelling on flat models. Three-dimensional models are also used. The stress state in such models is studied by the *method of freezing* (widely adopted) and the *method of diffused light* using special optical instruments.

In the freezing method special two-phase materials are used. The model is first loaded at a high temperature wherein most of the material softens and becomes non-resistive with the entire load borne by the skeleton (remaining solid portion in the material), which becomes deformed. The material is then cooled without being unloaded. The softened part solidifies, thereby freezing the skeleton. The model is subsequently sliced into strips. These strips are studied under a plane polariscope in the manner described above for flat models, in order to determine the quasi-principal normal stresses, that is, normal stresses in a given strip.

In the second method the loaded model is immersed in a liquid and a specific plane in the model examined using a thin strip of very bright plane polarised light. While observing this plane from one side, the picture of birefringence of light associated with the refraction coefficient and consequently with deformation at every point in the model is recorded. By directing a flat beam of polarised light in a different manner with reference to the model, the stress in any section of the model may be determined.

Just as in all other cases of modelling, while studying the physical phenomena associated with the stress state by the method of photoelasticity, it is required to maintain the similarity of corresponding criteria (complexes and simplexes) established either by the dimension method or by general differential equations describing the particular phenomenon.

For modelling the process of stress wave propagation in a blast, the similitude criteria of Euler, Cauchy and the Strouhal number must be met, as shown in [5, 6]. In modelling by the method of photoelasticity, as other materials are generally used, the conditions prescribed for the method of equivalent materials for studying elastic deformations should also be complied with. Modelling of elastic dynamic problems is further complicated by the polymer piezo-optic materials generally used, as these possess viscoelastic properties to some extent, which might considerably affect the intensity and pattern of wave propagation in the model, especially at high frequency vibrations and also in the presence of direct and reflected waves.

Currently, new materials from polymers, glass and other substances have been developed that possess piezo-optic properties with relatively low viscous characteristics. Their use ensures accurate similarity of stress fields. In certain cases investigations are conducted by the method of photoelastic coatings in which the model itself is fabricated from an opaque full-scale material while the coating material alone is transparent and possesses piezo-optic properties. However, in this case the distribution of deformation over the thickness of coating is associated with complex processes, due to which an uneven stress field is induced in the coating. As a result, errors arise whose magnitude and nature have not yet been studied adequately. The relationship between stresses in the object under study and the photoelastic coating is given by a relatively complex relationship [9]:

$$\sigma_{1;2} = \frac{E}{E_p(1-\nu^2)}[(1-\nu_p\nu)\sigma_{1p;2p}(\nu_p-\nu)\sigma_{2p;1p}].$$

To obtain the principal stresses σ_{1p} and σ_{2p} (principal deformations $\varepsilon_{1p}, \varepsilon_{2p}$) separately, the well-known methods in photomechanics are used, such as oblique (inclined) translucence of the coating, electric analogy, integration of equations of equilibrium pertaining to a planar problem and also the methods of measuring transverse deformations and absolute retardations by means of quantum optical generators (lasers).

It is necessary to fulfil the corresponding conditions of modelling and to use an optical-dynamic installation, while dynamically loading the model.

For example, to model the process of rock breakage by drills, the following conditions [9] should be fulfilled:

$$\frac{\Delta l_N}{\Delta l_M} = P_c l_c \frac{E_M\left(\sigma_{1N} - \dfrac{\sigma_{2N}+\sigma_{3N}}{\nu_N}\right)}{E_N\left(\sigma_{1M} - \dfrac{\sigma_{2M}+\sigma_{3M}}{\nu_M}\right)},$$

where Δl is deformation; l_c, P_c are geometric and force-related multipliers of similitude.

Here the similitude condition of contacts in the selecton of multipliers l_c and P_c is provided by selecting the model material according to the formula,

$$\left(\frac{E_{1N}}{E_{1M}} + \frac{E_{2N}}{E_{2M}}\right)^2 l_c = \left(1 + \frac{E_{1N}}{E_{2N}}\right) P_c.$$

Subscript 1 indicates the instrument, 2 indicates the material.

To model the process of blast-induced stress wave propagation, it is necessary to comply with the similitude criteria of Euler, the Strouhal number and the Cauchy criterion ($p/\rho v^2$; vt/l and $v/\sqrt{\varepsilon E/\rho}$), which amounts to the similarity in the method of equivalent materials for elastic deformations [5].

Transparent, isotropic materials which exhibit birefringence when subjected to external forces, are also used for optical investigations. Such materials are glass, celluloid, bakelite, xelolite, materials of gelatinous-glycerin base with betanapthal, epoxy resin with hardener (molenic or phthalic anhydrides) and various other additives.

Initially, specimens of these materials are subjected to calibration testing wherein the modulus of elasticity, Poisson ratio, stress-optic coefficient $C_\sigma = G/(\sigma, \delta)$ and $\tau_0^{1.0}$ are determined in the manner described above under uniaxial compression on the polariscope.

Investigations are often conducted on planar models or the picture of deformations of various flat sections is considered in the case of three-dimensional models.

The transition from plane strain in the selected sections to plane stress is made by the established formulae of the theory of elasticity:
Deformations in plane stress state

$$\varepsilon_{xx} = \frac{1}{E}(\sigma_{xx} - \nu\sigma_{yy}),$$
$$\varepsilon_{yy} = \frac{1}{E}(\sigma_{yy} - \nu\sigma_{xx}),$$
$$\varepsilon_{xy} = \frac{2(1-\nu)}{E}\sigma_{xy}; \qquad \ldots (3.27)$$

Plane strains

$$\varepsilon_{xx} = \frac{1}{E}[(1-\nu^2)\sigma_{xx} - \nu(1-\nu)\sigma_{yy}],$$
$$\varepsilon_{yy} = \frac{1}{E}[(1-\nu^2)\sigma_{yy} - \nu(1-\nu)\sigma_{xx}],$$
$$\varepsilon_{xy} = \frac{2(1+\nu)}{E}\sigma_{xy}, \qquad \ldots (3.28)$$

and elastic constants under the conditions of planar deformation

$$E' = E/(1-\nu^2); \; \nu' = \nu(1-\nu). \qquad \ldots (3.29)$$

In analysing the experimental results, the known relationships in the theory of plane wave propagation are used, according to which stresses σ_r in the direction of wave propagation and in a direction perpendicular to it σ_φ are determined by:

$$\sigma_r = -\frac{k^2}{4\pi}\left[\frac{1}{k^2c_P^2 r}f''\left(t-\frac{r}{c_P}\right) + \frac{4f'\left(t-\frac{r}{c_P}\right)}{c_P r^2} + \frac{4f\left(t-\frac{r}{c_P}\right)}{r^3}\right];$$

$$\sigma_\varphi = -\frac{k^2}{4\pi}\left[\frac{1-k^2}{k^2c_P^2}f''\left(t-\frac{r}{c_P}\right) - \frac{2f'\left(t-\frac{r}{c_P}\right)}{c_P r^2} - \frac{2}{r^3}f\left(t-\frac{r}{c_P}\right)\right], \qquad \ldots (3.30)$$

or, retaining the first terms in equations (3.30), we have approximately

$$\sigma_r = -\frac{1}{4\pi c_P^2 r}f''\left(t-\frac{r}{c_P}\right),$$
$$\sigma_\varphi = -\frac{1-2k^2}{4\pi c_P r}f''\left(t-\frac{r}{c_P}\right),$$

where $k = c_S/c_P$; c_P, c_S are velocity of propagation of longitudinal and transverse waves respectively; $f(t)$ function of pressure at source; r distance from explosive charge.

The non-stationary nature of loading processes on the model during a blast made it necessary to develop a special optical-dynamic polarising set-up (Fig. 9). It consists of a high-speed movie camera SFR, pulsating gas discharge lamps of various types (IFK-2000, ISSh-7, ISSh-400) or a pulsating laser as the light source, and special devices for simultaneous firing of charges (usually of lead azide) and synchronisation of blast, electric firing of lead wires or creation of a pulsating magnetic field, illumination, activating the SFR camera and oscillograph. The oscillograph is used for recording the parameters of stress waves or in some cases another method is employed (for example, piezotransducers or strain gauges), which is necessary for establishing and controlling the quantitative data. The pulsating laser PDU with an electronic-optical shutter enables one to obtain the laser radiation in the form of singular pulses or an entire series of pulses at a time interval $\Delta t = 0.045$ μs.

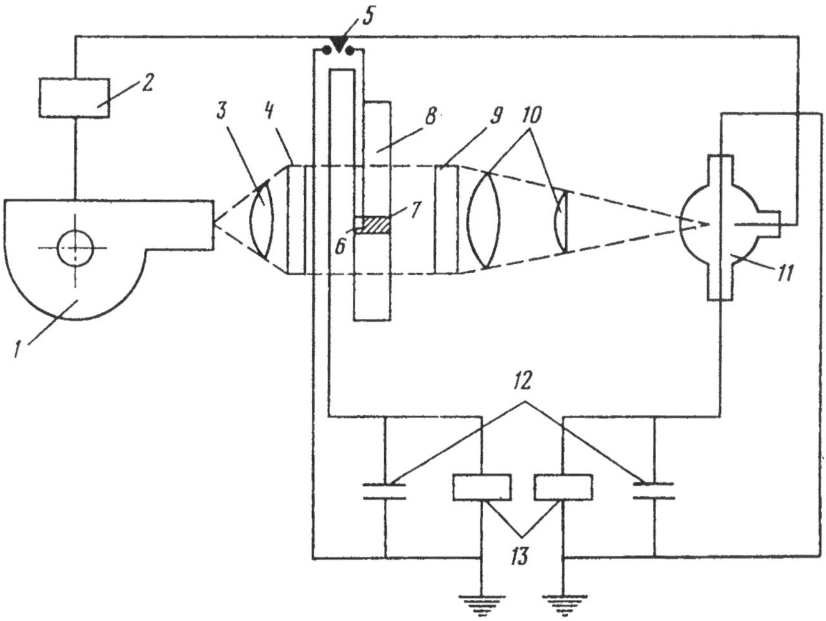

Fig. 19. Schematic diagram of an optically dynamic polarising installation.
1—movie camera SFR; 2—delay block; 3, 10—lenses; 4, 9—polaroids; 5—firing delay; 6—initiator; 7—charge; 8—model; 11—pulsating lamp; 12—pulsating capacitors for lamp and discharger; 13—charging device.

The stress state is recorded using the SFR camera in a regime of frame-wise snaps with a frequency up to $f = 2 \times 10^6$ 1/s while in the regime of photorecording with a series of different values of scanning speeds (for example, $v_{scan} = 750$ m/s and $v_{scan} = 3000$ m/s depending on the velocity of stress wave propagation).

The first regime is used for frame-wise visual recording of the picture of propagation and mutual interaction of stress waves at time interval (s) $\Delta t = 1/f$, while the second regime is used for continuous recording of the wave fronts under study and for determining the velocity of propagation as well as the intensity of waves (number of bands) in addition to the frame-wise snaps. Investigations are conducted using plane polarised light, and scattered (circulating) polarised light as well. The latter is required to remove isoclines from the isochromatic bands of the model to enable a more accurate depiction of them, in other words, for quantifying the value of the stress state.

Approximate similitude is adopted for investigating plane models and the following assumptions made: (1) stress state in the massif is studied within the framework of a plane problem posed in the theory of elasticity; (2) the effect of inelastic properties of rock on the stress state under study is neglected. Accuracy of approximation of the chosen sections in the plane models is determined by the condition.

$$\kappa = \delta\omega/c_P \ll 1 \quad \text{or} \quad \delta/\lambda_{dom} \ll 1, \qquad \ldots (3.31)$$

where δ is the thickness of the model; c_P the velocity of propagation of longitudinal wave, ω the angular frequency of the dynamic effect for stress wave $\omega \simeq \pi/(2\tau_N)$; τ_N duration of growth of wave amplitude to maximum; λ_{dom} predominant length of stress wave.

At $\delta \approx \lambda_{dom}$ and a distance $r \neq 10\delta$ the disturbance is transmitted uniformly over the entire thickness, with a velocity of c_P in the strip.

If the value of κ does not fully satisfy condition (3.31), a three-dimensional model may preferably be evaluated from the experimental errors in the planar models by the method of 'glued polariscope'. In this method the model is prepared from two polymer blocks, between which a thin strip of the same material is glued. Film type polaroid filters whose axes are oriented at 45° to each other are placed between the contact planes of the strip and the blocks.

Transition from the plane deformation in the chosen sections to the plane stressed state is made according to known formulae of the theory of elasticity. While deciphering the experimental results, the difference in normal stresses $(\sigma_r - \sigma_f)$ is established experimentally based on (3.24). The velocity of propagation of waves is computed based on the exposures of SFR in the recording regime, by the formula

$$c_{P,S} = (v_{scan}/M)\tan\varphi_{P\,S},$$

where v_{scan} is the scanning rate of the slot in the photorecorder; M image scale; $\varphi_{P,S}$ the angle of inclination of isochromes.

Generally, graphic interpretation of the stress state of a particular element by means of Mohr circles is used for determining the stress state in that element of model material. However, to determine the principal stresses one should also know the ratio between the principal stresses, i.e., σ_φ/σ_r. This ratio in the first approximation for elastic waves may be determined using relationship (3.26). This relationship can be reliably ascertained by experimental means, i.e., the direct measurement of the value of one of the principal stresses with the help of strain gauges or piezometers with simultaneous recording of differences in principal stresses by the photoelastic method on a dynamic optical-polarising device.

In conclusion, it is to be noted that the photoelastic method for modelling blast-induced phenomena in rocks, possesses all the disadvantages inherent in the method of equivalent materials. Moreover, the choice of equivalent materials meeting the requirements of optical penetration and optical anisotropy is also limited. Most such materials possess high viscosity and thus considerably distort the dynamic effects of a blast. Hence modelling by the photoelastic method may yield good results only for a study of the qualitative picture of wave interaction during simultaneous firing of a set of charges, by varying their mutual disposition, firing sequence and delay intervals.

3.1.5 Solution to Certain Problems Accompanying the Blast Effect in Rocks, by the Photoelastic Method

As a first example let us consider the possibility of modelling stress waves arising out of blasting two extended charges, by the method of dynamic photoelasticity. Under dynamic loading of the model, application of relationships (3.24) and (3.26) requires additional substantiation, as the physicomechanical properties of transparent isotropic materials used in modelling depend largely on the parameters of loading. Experience has shown that certain materials of the type acrylic plastic (PMMA), hard epoxy compound (EKZh) enable use of such relationships under dynamic loading conditions, with error not exceeding 10–15%.

The non-stationary nature of loading processes on the model in a blast demands use of the special optico-dynamic polarisation device described above. The mutual interaction of opposing stress waves induced by firing elongated charges placed in two neighbouring parallel shot holes was investigated. The problem involves planar symmetry relative to the planes parallel to axes of shot holes and passing through the axes of shot holes as well as in the middle of them. Thus the investigations may be confined within two sections: first, by the plane of shothole location (model of the second type) and second, by the plane perpendicular to the axes of shot holes (model of the first type). Models were prepared from acrylic plastic sheet (PMMA) and hard epoxy compound (EKZh) with a thickness $\delta = 3$ to 5 mm. Experiments were conducted at varying distances L between axes of shot holes. Model dimensions were such that waves reflected from their boundaries exerted no perceivable effect on the zone under investigation.

The interaction of opposing waves, generated by firing *EC* in shot holes, and the propagation of cylindrical wave were studied on flat models of the first and second type. For processing the results, the known relationships from the theory of propagation of plane waves [(see formulae (3.27) to (3.29)] were used. Models of the second type were loaded by the parallel and loop type firing of shothole charges of desensitised TEN. The modelling results thus obtained could be transferred to the full-scale within the framework of the adopted scheme and based on formulae related to the similitude of dynamic problems for elastic materials. The diagrams of radial component of longitudinal and transverse stress waves were plotted by processing the experimental results.

Diagrams of radial component of stresses $(\sigma_r)_{xy}$ of a cylindrical wave in a strip, obtained for two different distances, are shown in Fig. 20. The dynamic Poisson ratio in the computations of these diagrams was $\nu_{dyn} = 0.25$. The transverse wave occurring in the blast-loading of the strip was observed in the photographs at distances $r \geqslant 30R_{0_z}$ in the model of the first type as a set of isochromes progressing with velocity c_S. The transverse wave was also noticed in the diagram of isoclines on movie pictures obtained in the continuous recording regime and filmed in plane polarised light. The occurrence of transverse waves under symmetrical loading, is probably associated with the action of moving forces along the contour of the cylindrical charge cavity when the detonation wave of the *EC* interacts with it.

Interaction of two opposing cylindrical waves in the strip (model of the first type) is represented on the film pictures obtained at a distance $r = 50R_{0_z}$. Analysis revealed that up to the moment of encounter, each wave progressed at a velocity of $c_P = 2440$ m/s. The stress state characterising

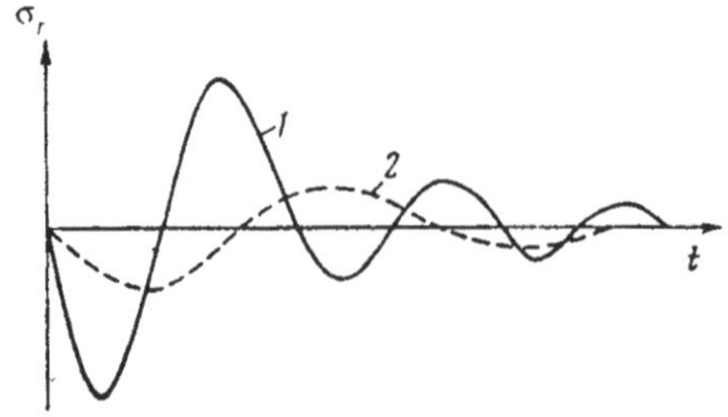

Fig. 20. Diagrams of the radial component of a stress wave in a strip at distances of (1) 20 R_{0_z} and (2) 40 R_{0_z}.

the interference of initial phases of longitudinal waves, developed after the progressive fronts met. This is shown schematically in Fig. 21, where the relative positions of the diagrams of longitudinal interacting opposing waves over time are shown. Notations $ab, a'b'$ and $bc. b'c'$ in the figure are corresponding initial phases of compression and tension; $cd, c'd'$ and $de, d'e'$ are phases of secondary compressions and tensions.

Here dimensionless time $T = 2c_P/\lambda t^*$ (where λ is length of compression phase; t^* time reckoned from the moment of contact of progressive fronts of longitudinal waves) was adopted. The moment of meeting of waves is shown in Fig. 21, A, while the moment of total interference of initial phases of compression of longitudinal waves is shown in Fig. 21, B. Fig. 21, C corresponds to the mutual effect of initial phases of compression in the waves followed by initial phases of tension. The complete interference of initial phases of

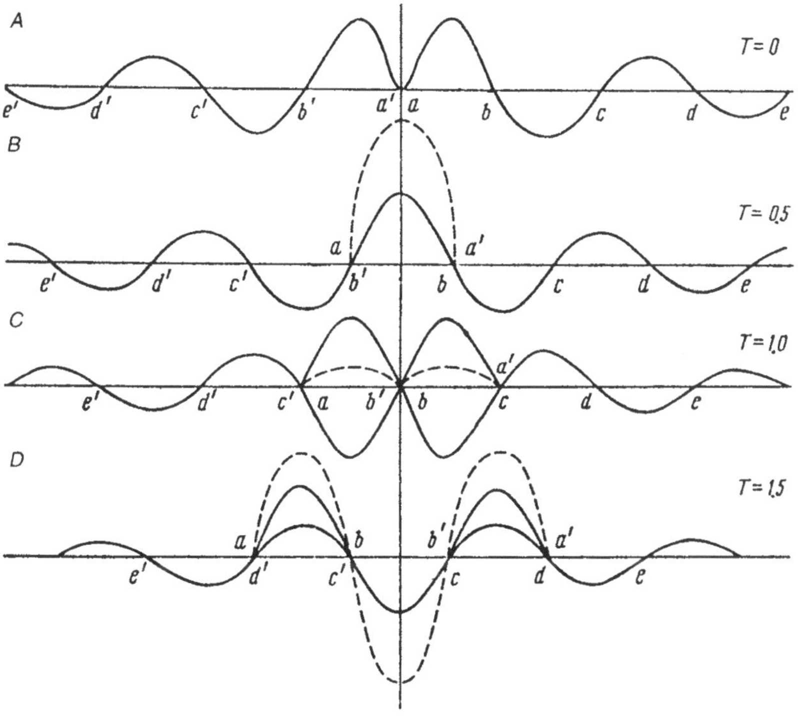

Fig. 21. Schematic diagram of interaction of opposing waves depending on relative time T. A—moment of meeting of waves; B—moment of total interference of initial phases of compression of longitudinal waves; C—mutual effect of initial phases of compression in waves followed by initial phases of tension; D—complete interference of initial phases of tension.

tension is shown in Fig. 21, D. At this time period the initial phases of compression of waves interact with the secondary phases of compression. The process continues in a similar manner later.

It can be seen from Fig. 21 that maximum stresses arise when the initial phases of compression and tension of opposing waves interact. The transverse wave, progressing at a lower velocity, is situated in the tail of the longitudinal wave. Intensity of transverse waves was not very high; hence their interference did not develop significant stresses in the zone of wave interaction under study.

If the distance between the charge axes is reduced, the interference of transverse waves may considerably affect the stress state. This is because at the moment of interaction of longitudinal waves at such distances, transverse waves do not succeed in separating from them and hence also influence the interference pattern. At a distance of $r = 20R_{0_z}$, transverse waves contribute 22% to the amplitude value compared to the amplitude of stresses in the compression phase of a longitudinal wave.

Let us consider the nature of breakage based on the variation in radial σ_r^{Σ} and tangential $\sigma_{\varphi}^{\Sigma}$ components of the sum total stresses along the line of shot holes and perpendicular to it, depending on the dimensionless time of interference $T = \dfrac{2c_P}{\lambda}t^*$ and spacing between shot holes L. The experimental data for σ_r^{Σ} and $\sigma_{\varphi}^{\Sigma}$ is given in Table 2 for relative distances between shot holes $\bar{L} = L/R_{0_z}$ and for points 1, 2, 3, 4 (lying on the x-axis) and 6, 7, 8 (lying on the y-axis) lagging uniformly from the origin (x, y). The origin is taken to be on the line of shot holes at their centre; the x-axis is directed along the line of shot holes and the y-axis perpendicular to it. Spacing between points is taken to be $\Delta L = \lambda/4$, i.e., one-fourth the length of the longitudinal wave, which in the experiments varied within the limits of given spacing between shot holes $L = (20 \text{ to } 50)R_{0_z}$.

Analysis of the tabulated data (Table 2) shows that just as in the process of wave propagation, mutual alternating (varying sign) loading of material in the model along the line of x and y axes occurred, even in the case of mutual penetration of waves. The nature of loading induced by both the components of wave σ_r and σ_{φ} exhibits in most cases commensurate amplitudes in both the compressive and tensile phases of the wave diagrams. The fronts of longitudinal waves began to meet on the axis of symmetry between shot holes and subsequently spread into this entire area.

The stress state in the interference area for subsequent similar time periods, as mentioned above, moved in the direction of propagation of detonation with a velocity $v = c_P/\sin\alpha_1$. The qualitative picture of interference of opposing waves did not change when the axes of shot holes were brought closer. The nature of the diagram changes to some extent and the amplitude values of summed stresses increase, while the nature of loading due

Table 2

L	T	Stress values σ_r (in the numerator) and σ_φ (in the denominator), 10^{-5} Pa, at points						
		1	2	3	4	6	7	8
	0	0/0	−995/−288	0/0	+550/+160	0/0	0/0	0/0
	0.5	−1200/−350	0/0	+510/+110	0/0	−1170/−510	−1100/−500	−770/−400
50	1.0	0/0	−90/−25	0/0	−225/−73	0/0	0/0	−720/−280
	1.5	800/233	0/0	−780/−230	+100/+30	+775/+350	735/340	515/270
	2.0	0/0	300/58	−150/45	−40/−115	0/0	150/55	600/120
	2.5	−220/−83	90/38	450/185	0/0	−210/−88	−230/−85	140/−68
	0	0/0	−840/−320	0/0	878*/485*	0/0	0/0	0/0
	0.5	−1700/−615	0/0	950*/340*	136*/136*	−1550/−750	−600/−325	0/0
30	1.0	700/255	0/0	−70*/70*	−330*/−75*	−250/−90	−340/−75	−1550/−915
	1.5	1650/600	0/0	−530*/−35*	85*/35*	1500/725	585/310	120/85
	2.0	35*/35*	205*/77*	0/0	−870*/−140*	125/50	375/150	750/470
	2.5	−1030*/−38*	0/0	692*/68*	0/0	−690/−333	−315/−90	−55/−40
	3.0	25*/30*	−180*/175*	165*/−115*	375*/135*	−23/−20	−780/−310	−175/−405
	0	0/0	−1200/−530	−105*/−105*	— —	0/0	0/0	0/0
	0.5	−2400/−1060	−19*/18*	660*/470*	— —	−1100/−910	−800/−525	0/0
20	1.0	0/0	660*/465*	0/0	— —	0/0	−1200/−500	−1075/−625
	1.5	1400*/535*	0/0	−35*/35*	— —	750*/325*	0/0	−1400/−225
	2.0	220*/485*	350*/140*	140*/140*	— —	305*/400*	150*/200*	75/30
	2.5	−840*/−350	175*/35*	475*/320*	— —	−720*/−315*	−280*/−280*	110*/135*
	3.0	−125*/−50*	−100*/−48*	0/0	— —	−205*/−60*	800*/−140*	−410*/−160*

NOTES: 1. The (+) sign denotes tensile stresses and (−) compressive stresses.
2. Numbers with a star indicate the summed value of stresses due to interference of longitudinal and transverse waves.

to σ_x and σ_y coincides sign-wise in most cases, i.e., either compression (−) or tension (+). Compression of material is observed initially, followed by tension etc., and the first cycle is repeated again.

Visual observation of the models after experimentation showed that the sections along the line of shot holes (along the x-axis) and along the line of symmetry (y-axis) happen to be the most intensely loaded zones in the model.

At low relative distances between shot holes ($\bar{L} \leqslant 30$), the occurrence of through-cracks between shot holes and also cracks along the line of symmetry were observed, which in some cases reached the edges of the model in the direction of LLR.

At large relative distances ($\bar{L} > 30$), cracks between shot holes remained closed. However, these were very long compared to the remaining radial cracks. In this case cracks were not observed even along the axis of symmetry of the model.

Crack formation between shot holes is caused mainly by the action of tensile forces arising from the tangential component σ_φ, as these are directed perpendicular to the line joining the shot holes. In the formation of cracks along the axis of symmetry at low relative spacing between shot holes, evidently a key role is played by the high tensile stresses due to the bending of strips.

Actually, at a relative spacing between shot holes (see Table 2) $\bar{L} < 30$, these stresses exceed the critical strength of acrylic plastic $(\sigma^{dn}_{rup}) = 50$ to 65 MPa, at all points situated both on the axis of symmetry and the line of shot holes.

The interaction between two opposing waves in the plane of shot hole axes (model of the second type) is shown in Fig. 22 in the form of diagrams of wavefronts generated by firing two adjacent charges, separated by a distance $L = 25R_{0_z}$. The fronts of longitudinal and transverse waves in the middle portion of the strip are conical in shape and are situated at certain angles to the direction of axial lines of shot holes, quantified by the ratio of propagation velocities of these waves and velocity of detonation D:

$$\sin\alpha_1 = c_P/D;\ \sin\alpha_2 = c_S/D.$$

The schematic location of the fronts of longitudinal and transverse waves until the moment they meet, when two adjacent charges are directly initiated and both are simultaneously blasted, is shown in Fig. 22, a. The interference pattern of waves along the axis of symmetry (axis y) moves in the direction of detonation with velocity

$$v_f = c_P \sin\alpha_1.$$

The wave pattern obtained under the simultaneous meeting-type initiation of two shot hole charges situated at the same distances as in the previous case, is given in Fig. 22, b. Here, a similar nature of the location

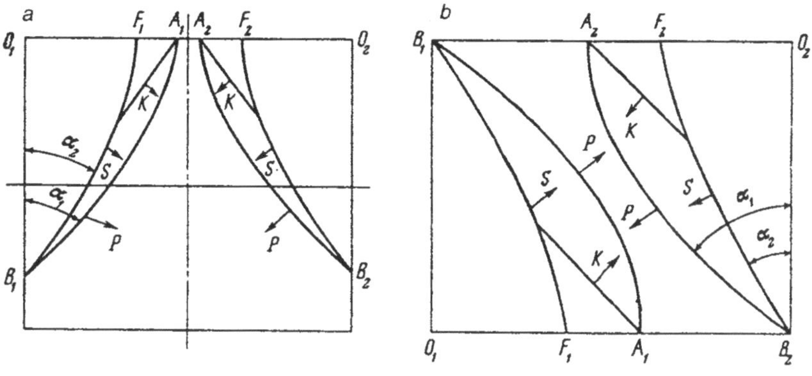

Fig. 22. Mutual interaction of two opposing waves, in a plane passing through the axis of shot holes.
A_1, A_2—meeting points of longitudinal P and conical K waves; B_1, B_2—meeting points of longitudinal P and transverse S waves; O_1, O_2—points of initiation of charges 1 and 2 in adjacent shot holes; F_1, F_2—intersection of transverse wave front with upper or lower edge of block.

of fronts of longitudinal and transverse waves was observed. The direction of their propagation only altered due to meeting-type initiation of charges.

The stress state under this type of loading by direct initiation, is distinguished by a different sequence of wave interference. As a result, maximal stresses arising at the moment of total interference, are realised in the zone situated on the axis of symmetry, roughly at the middle of the shot hole length.

As a result of the model experiments conducted, a qualitative pattern of interaction of opposing stress fields was established for a system consisting of two shot hole charges; the wave pattern in the main planes of the loaded model was obtained; and stresses arising in the zone of interference and the nature of breakage based on the distance between sources of disturbance were evaluated. When the charges were placed closer ($L \leqslant 20R_{0_z}$), the transverse wave also influenced the sum total value of stresses in the case of wave interference, by enhancing it considerably. Consequently, the spacing between charges $L = 20R_{0_z}$ may be assumed as the boundary of effective utilisation of pair-wise closely located charges. It was established that meeting-type initiation ensures uniform loading of the object being blasted with a characteristic break along the line of location of charges, caused by the action of the tangential component of stress wave σ_φ, and this may be successfully used for the purpose of obtaining improved profiles of bench highwalls and in contour blasting.

The *second illustration of the use of photoelasticity* in model experiments serves for evaluation of dynamic strength of rocks at the slabbing stage, by the method of optically sensitive coatings on an optical-dynamic device, which happens to be a reflecting polariscope with 20° angle of rotation (Fig. 23). Deformations in specimens are established through movie pictures of isochromes filmed in polarised light with high-speed camera SFR in the photorecording regime at a speed of 750 m/s. Strips of epoxy compound 10 × 70 × 2 mm were used as coatings and glued to the specimens by an epoxy gum of the cold-curing type. The contact side of the strip was covered with a thin metallic-mirror layer.

Values of maximum tensile strains ε_{max} and speed of pulse progression in the rod c_{pulse} were determined by the method of coatings. Based on these, the dynamic modulus of elasticity was computed, $E_d = \rho_p c_{pulse}$ (where ρ_p is the unit weight of rock). Knowing the values of E_d and ε_{max}, the maximum tensile stresses were determined, $\sigma_{max} = E_d \varepsilon_{max}$. Error in the computation of ε_{max} did not exceed 9% and that in σ_{pulse} 5%.

Dynamic breaking strength was determined by using rod-shaped specimens 10 × 10 × 100 mm, loaded by firing microcharges of *EC* weighing 0.05 to 0.16 g, strapped to one side.

The results obtained characterise the strength of rock with unflawed structure under high-frequency pulse action of rarefied wave reflected from the other end of the rod, over a duration up to 10^{-5} s. Usual breakage in the form of slabbing or fracture occurred at a distance of 15–25 mm from the end opposite to the site of loading. These breakages were induced by the action of a tensile pulse formed due to reflection from the free end of an initial pulse of compression generated by firing the microcharge.

Using the method described above, the probable mode of breakage of specimens under different values of pulses of stress wave, induced by firing charges of different weights, was determined. The experiments involved charges of the same weight being blasted 15–20 times.

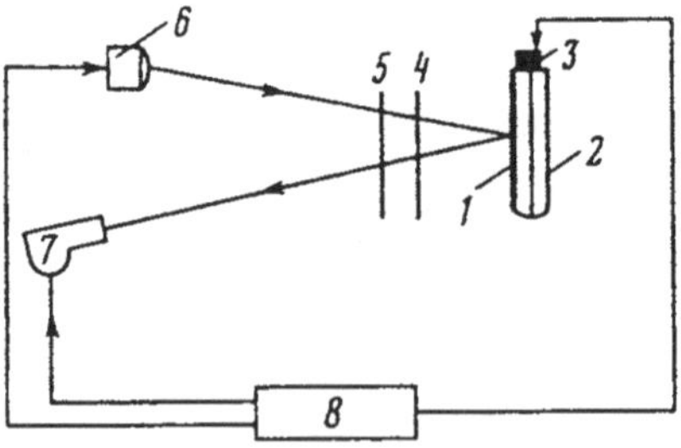

Fig. 23. Schematic diagram of an optico-dynamic device in the form of a V-shaped reflecting polariscope, used in the method of photoelasticity in the variant of optically sensitive coatings. 1—optically sensitive coating; 2—sample; 3—charge; 4, 5—polaroids; 6—pulsating lamp; 7—SFR camera; 8—control panel.

The results of these model experiments are shown in Fig. 24, where the probabilistic relationship of dynamic breaking strength of two types of granite is presented: in Fig. 24, a, a hard, weakly fractured granite with $\rho_p = 2630$ kg/m^3; $c_P = 4210 + 150$ m/s and $\sigma_{rup}^{st} = 12.5$ MPa; in Fig. 24, b, a granite of medium hardness, heavily fractured with $\rho_p \approx 2610$ kg/m^3, $c_P = 4000 \pm 150$ m/s and $\sigma_{rup}^{st} = 11$ MPa. Each of the granite types was tested under room conditions (curve 1) and also under conditions modelling the temperature-humidity regime in permafrost (curve 2). As can be seen from Fig. 24, the dynamic strength of permafrost granites is higher than that of ordinary granites by 11%, while the dynamic tensile strength under 50% probability of breakage is equal in the first case (see Fig. 24, a), 68 and 80 MPa respectively, and in the second case (see Fig. 24, b), 60 and 70 MPa. At 20% probability of breakage, the dynamic tensile strength would be respectively equal to 40 to 55 MPa in the first case, and 35 to 45 MPa in the second. The latter data has been confirmed by experimental results of dynamic tensile strength for the granities obtained in full-scale conditions when powerful charges with radial air gap in blast holes of various diameters (varying gap sizes) were fired [4]. Therefore data with 20% probability happen to be more dependable for determining the dynamic strength of rocks by the method described above.

3.2 Modelling Blast Effects Based on Dynamic and Energetic Similitudes under Laboratory and Full-scale Conditions

In the modelling of various phenomena arising due to a blast in rocks, certain methods are widely used in which either rock specimens of different shapes or the rock massif itself are used as the model. For the first model experiments are usually conducted under laboratory conditions by using various

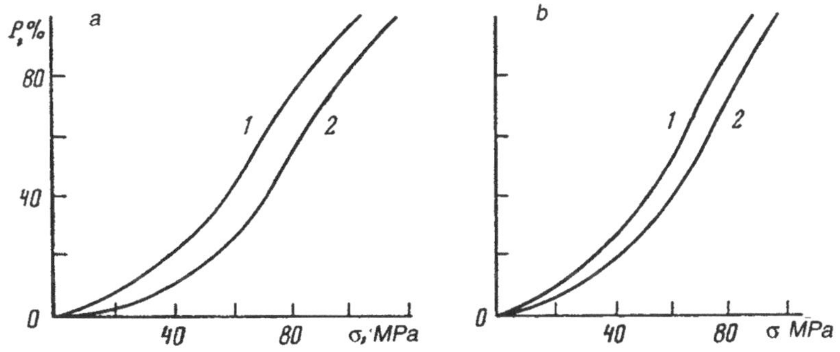

Fig. 24. Probability versus dynamic breaking strength.

installations and methods of recording parameters of the phenomena and processes under study, while in the second the rock massif per se is usually subjected to charges of reduced weight under special conditions of blasting, using single or group charges.

The processes of progression of explosion in the near-field zone (for example, the development of a cavity under the effect of gaseous products of blast, process of formation and development of compression wave) and in the mid-field zone (for example, development of zones of pulverisation and crack formation, patterns of stress wave propagation, processes of interaction between adjacent charges, reflection of stress wave from the free face, interaction between incident and reflected stress wave etc.), as well as the processes of breakage and crushing of specimens under different conditions of blasting and a series of other phenomena (effect of stress wave on mine workings or effect of blast-induced seismic waves on surface and underground structures) are studied in laboratory conditions. Depending on the objectives and tasks of the experiment, various devices and instrumentation are used based on the methodology of conducting the model experiment. These devices are specifically developed and fabricated in the process of setting up each experiment.

Experiments for studying the parameters and patterns of propagation of air blasts (AB) and blast-induced seismic waves (BSW), arising out of firing single and group charges under different conditions of blasting as well as the processes of action of such waves on buildings and engineering structures, are studied under conditions of a full-scale massif. In such conditions it is also advisable to investigate other features of blasting, such as fly rock, rock mass displacement and formation of muck pile. It should be noted that these features are also studied in industrial full-scale blasts if the condition of synchronisation of the blast with triggering of the measuring devices is satisfied.

Different methods of conducting model experiments and the necessary modelling conditions for them in the case of sequential study of the processes of blast effects in rock, starting from phenomena occurring in the near-field zone of blast and ending with the effect of a blast on buildings and engineering structures, as well as modelling the process of rock fragmentation are presented below.

3.2.1 Methods of Recording Processes Occurring in Near-field Zone of a Blast

Among the various phenomena occurring in the near-field zone of a blast, the following are of paramount interest: refraction of detonation wave in rock, resulting in the formation of a compressional wave with a shock front in rock; progression of shock wave into the rock, followed by break-up of the shock front and its transformation into a stress wave; expansion of charge column (camouflet cavity of blast) under the effect of gaseous products of a blast, creating a quasi-static stress field in the surrounding rock. Recording these rapidly occurring phenomena under very heavy pressures, varying from 10^8 to 2×10^{10} Pa, by means of various transducers (piezotransducers, strain gauges etc.) placed near charges of cylindrical and spherical shape, is a highly complex and very unreliable task. Apart from the potential damage to the equipment, severe disturbances occur which are associated with electromagnetic radiation, influence of dimensions of the transducers and also the methods of their location and fixture in the rock near the charge. Hence in pressure-time records electromagnetic disturbances are recorded as precursors while the wave pressure per se is recorded right up to that pressure under which the transducer is damaged. Thus in a favourable case records of pressure up to the maximum value of wave amplitude (manganin transducers) are obtained. But the shape of the pressure-time diagram may be distorted by the dimensions of the transducer.

In such cases the more dependable methods for recording the phenomena listed above are photographing movement in the chamber wall and outer edge of the specimen with a high-speed movie camera, for example SFR-2M, or recording X-ray pulses. In the *first case,* cinematography, it is possible to record not only movement of the chamber wall and outer edge of the specimen, but also movement of the compressional wavefront in specimens made of optically transparent materials such as glass, bakelite etc. In the *second case,* X-ray pulse recording, it is possible to record movement of the chamber wall and outer edge of the rock specimens. The thickness of the specimens would have to be limited by the penetration capacity of the X-rays.

The shadow device IAB-451 in combination with high-speed movie camera SFR-2M is used for optical motion picture recording.

The optical scheme of shadow device IAB-451 with movie camera SFR-2M is shown in Fig. 25. It consists of two optical systems: a collimator system, emitting parallel beams of light to illuminate the test specimen, and an observational system that enables observation and photography of the shadow of the image.

The collimator part consists of a reflecting meniscus lens (4) of 230 mm diameter with a focussing distance of 1917 mm, a reflecting prism (3) for altering the direction of rays, fixed on one of the frames with a slit (2) and placed in the focal plane of lens (4). The frame is equipped with a mechanism for rotation and changing of slit width. The observational system comprises a reflecting meniscus lens (7) similar to (4), reflecting prism (8) fixed on a frame having a stem (9) and installed in the focal plane of the lens (7). Lens (10) may be moved along the optical axis for a sharp picture of the objects lying on plane S–S on the screen (film tape (12)). Variation in slit width and movement of the stem are achieved by micrometer screws

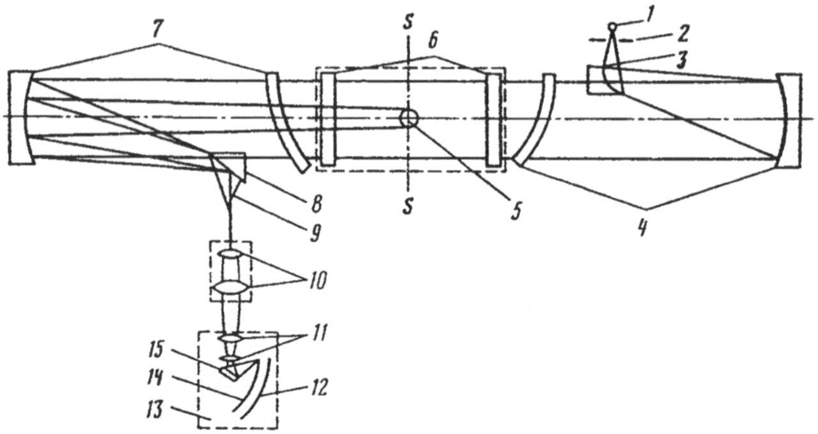

Fig. 25. Schematic diagram of shadow device IAB-451 with movie camera SFR-2M.

with an accuracy up to one-hundredth of a millimeter. The height of the slit may be altered with an insert diaphragm. The slit is evenly illuminated by the light source (1) by means of a condenser. Light passes through the slit (2) to the reflecting prism (3), falls on the mirror of lens (4) and on being reflected passes to the meniscus of lens (4) in the form of a parallel beam of light, through the specimen (5) and the mirror-meniscus lens (7), and being reflected here falls on the reflecting prism (8). Now the beam, after being reflected from (8) and passing through lens (10), falls on the recorder chamber (13). As the reflecting meniscus lenses (4) and (7) are completely identical, the dimensions of image of the slit or object in the S–S plane are equal to the dimensions of the slit or object itself. Protective glasses (6) safeguard lenses (4) and (7) from blast effects.

When a charge of *EC* is fired, as a result of intense disturbances initially in the homogeneous, optically transparent medium (air, water, glass etc.) within the field of tube IAB-451, deflections of light beams occur due to the formation of surfaces of higher gradient of density (shock front, surface of expanding cavity). Due to this the optical disturbance projected on the screen appears to be less illuminated than the undisturbed part of the medium. A system of light bands appear on the screen whose configuration corresponds to that of sudden jumps in shock waves of the explosion products and the surface of the cavity.

A recording method based on a rotating mirror system appears to be the best method for studying transient processes having a life not exceeding several microseconds. In this case the image can move over an immobile film at high speed. The SFR camera is used for recording. The frequency of mirror rotation varies from 3×10^3 up to 75×10^3 revolutions/min at a linear scanning speed ranging from 150 to 3750 m/s. The SFR camera may also be used as a photorecorder for obtaining continuous scanning of the process over time and also as a time loop variant for obtaining sequential photographs of the process under study with a frequency of filming 2.5×10^4 to 2.5×10^6 frames/second (frame-wise recording).

In the time loop variant the image of object (5) is constructed by two lenses (11) in a plane intersecting the centre of the reflecting rib of the mirror (15). At a certain distance from it, along an arc of the circle, several square-shaped lenses assembled in the form of an insert lens (14) are placed. Before the insert lens a diaphragm D_2 is also placed on the arc of the circle, and before the first lens (11) four-row or two-row square diaphragms D_1 are placed, which have correspondingly four slits shifted horizontally and vertically or two slits. The system of diaphragms and lenses acts like an optical shutter, whereby a series of separate sequential photographs of the process can be obtained on an immobile film (12). Synchronising the illumination of the process with the moment of initiation and the beginning of mirror rotation is achieved by a specially developed synchronising device that facilitates charge initiation at any position of the mirror and recording the process right from the moment of its development.

A movie camera in the variant of photorecording enables study of the occurrence of the process in a particular section of the model field, formed by the node of the slit of the movie camera, which can be varied from 0.1 to 0.5 mm in width. In this case the insert lens with diaphragms D_1 and D_2 (optical shutter) is absent, while the image section is moved by the rotating mirror along the focal surface where the immobile film tape is fixed. Thus the process under study is recorded as a continuous video tape over a time period. Here the scanning speed is controlled by the speed of mirror rotation, which gives a good resolution over time. Scanning speed ($v_{scan} = 250$ to 3750 m/s)

$$v_{scan} = bv_f/(u_f M),$$

where b is the width of diaphragm and slit node; u_f the permissible slurring of image on the tape; M scale of recording; v_f rate of progression of the process under study.

In frame-wise recording it is possible to investigate development of the expanding crater cavity, the front of a shock or stress wave, interaction of stress waves and a series of other events associated with cavity expansion and stress waves. In photorecording the wavefronts are tracked precisely and the speed of progression of the wavefront, when inclined at an angle α to the horizontal plane, is given by:

$$v_f = v_{scan}\frac{\tan\alpha}{M}$$

where v_{scan} is the scanning speed of the photorecorder.

The method of X-ray recording allows recording of the deformation process in an opaque media and development of the cavity over time, whenever adoption of other experimental techniques would prove difficult.

The former USSR was the first country to use the X-ray graphic method for investigation of formation of crater cavities created by blasting charges of *EC* in media possessing mechanical strength. Furthermore, A.N. Khanukaev was the first to apply this method to problems associated with rock breakage.

The X-ray pulse apparatus (Fig. 26) consists of generators of pulse voltage (GPV), X-ray tubes (XT), vacuum system for creating and supporting in them a high degree of vacuum, delay circuit for operation of X-ray tubes, charging circuits and devices and other auxiliary nodes [7]. The GPV, after being charged by a rectifier, generates pulses of 800 kV which are supplied to the electrodes of the X-ray tubes. At the initial stage of vacuum breakdown in the interelectrode interval, a very short-lived (0.15–0.25 μs) X-ray radiation arises. Due to the presence of a delay circuit in the discharge surface of the GPV, the X-ray tubes operate sequentially with microsecond intervals. No tube emits for more than 2×10^{-7} s. The delay lines in the discharge surface provide the delay for operation of each tube relative to the preceding one by 2 to 7 μs. The scheme of measurement of time intervals between the periods of operation of tubes ensures the formation and recording of points on the oscillogram corresponding to each X-ray flash.

The test specimen is placed in the middle of the blast chamber, around which along the 90° sector, sharp-focussing X-ray tubes are installed. Cassettes with X-ray film are placed in the blast chamber exactly opposite to the specimen to be blasted and the X-ray tubes. The film is then exposed.

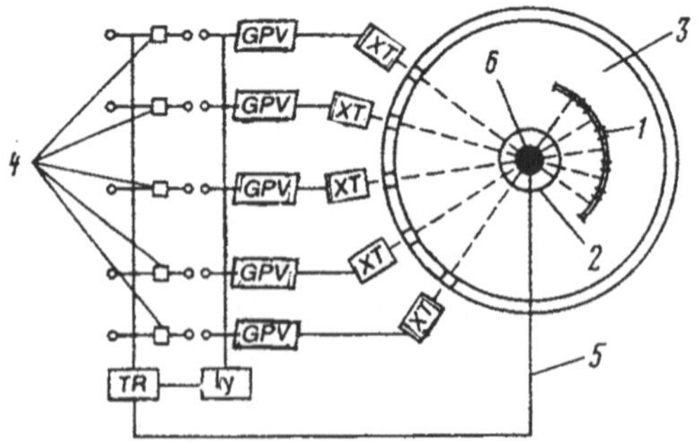

Fig. 26. Schematic diagram of X-ray pulse apparatus.

1—X-ray film cassettes; 2—test model; 3—blast chamber; 4—delay line; 5—firing cable; 6—EC charge.

Deciphering X-ray pictures resolves into the measurement of cavity radius on the prints and recording of time intervals corresponding to increments in cavity radius. In so doing the coefficient of geometric enlargement of images, equal to 1.15 under experimental conditions, is taken into account.

Rock specimens of spherical or cylindrical shapes 35–55 mm in diameter are used as the model. The explosive charge is placed in holes at the centre of a spherical specimen or along the axis of a cylindrical one. To obtain high contrast X-ray pictures of the blast, the internal surface of the cavity is lined by lead foil 0.1 mm in thickness. The X-ray apparatus is triggered by the closure of an ionising sensor situated at a particular distance from the point of initiation of charge (drop of lead azide). Knowing this distance and the detonation velocity of the explosive, we can determine the delay time in initiating the entire charge which is actually considered for synchronising the instant of blast with that of beaming of X-ray radiation (operation of X-ray tubes) and starting the oscillograph OK-17M. The OK-17M has two channels. First channel has calibrations corresponding to each X-ray flash, while the second channel has time calibrations arising from the time signal generator.* The discharge pulse of current is drawn through resistance R_{dis} and is supplied via the formation device to the vertical deflecting plates of the OK-17M oscillograph [2].

To investigate the above phenomena, the modelling conditions arising out of the extended law of geometric similitude are used, which are based on the energetic similitude in blasts [5, 6]. Additional modelling conditions are imposed by the problems of a given investigation.

3.2.2 Methods for Conducting Model Experiments for Studying the Processes Occurring in Mid-field Zone of a Blast

Various methods are currently available for determining the dimensions of zones of compression, breakage or fragmentation and crack formation, as well as the parameters of stress and seismic waves induced by a blast in rock through model experiments. The common feature amongst them is the adoption of blasts of reduced charges, even up to very small sizes (microcharges), compared to full-scale blasts. Blasts are conducted in full-scale media, either in monolithic sections of the natural massif or in blocks of various sizes cut from natural rocks, and also those made from optically transparent materials (in laboratory conditions). To model such blasts, the extended law of geometric similitude is applied wherein the ratio of dimensions of charges in the full-scale model, equivalent with regard to explosion

---

* The first channel records the short time history while the second records the long time history of the process—Translator.

energy (usually radii or equivalent square roots of charge masses proportional to them, i.e., $\sqrt[3]{G_*}$ or $\sqrt{q_*}$) is taken as the linear scale of modelling and the media of model and full-scale are assumed to be similar.

All such methods are distinguished mainly by the methodology of experimentation and measuring apparatus used for determining the parameters of blast phenomena under consideration.

Usually, rock blocks of various sizes and microcharges for firing them are used for determining breakage zones. To record unknown parameters, various instruments are used, starting from simple ones such as visual observation and photography up through very sophisticated apparatus of optical recording using high-speed movie cameras and special electronic devices for synchronising the working of this apparatus with the timing of detonation.

To investigate the stress wave parameters in the mid-field zone of blast, the so-called *method of refracted wave* is perhaps the most reliable. The salient feature of this method is that stress in rock is not measured directly; instead, the pressure of refracted wave in water is measured by piezo pressure gauges, installed near walls of the block in water. Such an approach is adopted because direct measurement by similar gauges placed in rock is associated with a set of factors leading to large errors in measurement. The dimensions of the gauge and the hole drilled for placing it exert an influence on the stress field to be measured, which in turn contributes primarily to such errors. Furthermore, accurate positioning of gauges in relation to the explosion and reliable contact between gauge and rock also pose difficulties. Usage of various binding materials, in this case cement, plaster of paris etc., also introduces certain errors in measurement, as the acoustic resistance of such materials always differs from the acoustic stiffness of the surrounding rock, while the gauge contact with the rock is not always stable. This means that direct measurement by a transducer positioned in the rock may give only a comparative evaluation of the stress field. This can be quantitatively evaluated by the method of refracted wave.

According to this method, the pressure wave refracted from the rock into water is measured by a piezoelectric pressure transducer and special electronic apparatus. The parameters of the stress wave at the wall of the block are established by computations, using the laws of refraction of elastic waves based on acoustic approximation formulae:

$$\sigma = p/k_{ref}, k_{ref} = (2\rho_w c_w)/(\rho_w c_w + \rho_p c_p), \qquad \ldots (3.32)$$

where p is the pressure measured by the transducer; k_{ref} the refraction coefficient; ρ_w, c_w density and velocity of propagation of sound in water; ρ_p, c_p density and velocity of propagation of sound in a block of rock. Based on these formulae the normal component of stress wave (normal to the plane of refraction, which means the wall of the block) is determined for a blast with a concentrated charge. In the case of blasting a concentrated or spherical charge, it is determined by the relationship.

$$\sigma_p = \sigma_r \cos^2 \alpha + \sigma_\varphi \sin^2 \alpha, \qquad \ldots (3.33)$$

where σ_r is the radial component of stress wave; $\sigma_\varphi = \sigma_\theta$ is the tangential component of stress wave; α the angle formed between the vector $\bar{r}$ drawn from the blast centre to the observation point and the normal to the block wall at the points of refraction (observation).

Evidently, at $\alpha = 0$, i.e., if the vector $\bar{r}$ coincides with the normal to the observation point, then only the radial component of stress wave σ_r, i.e., normal component to the spherical stress wave front, would be recorded.

This method was initially developed for measuring waves induced by concentrated charges alone, with certain assumptions as detailed below.

In formulae (3.32) the values of ρ_w, c_w, ρ_p and c_p were taken for undisturbed media, which is justified for waves of small amplitude (seismic and acoustic waves). Actually, we should take the values of ρ_p and c_p for points of maximum amplitude of stress. Error due to this assumption does not exceed 8% while measuring the stress waves starting from distances of $r = (10 \text{ to } 15)R_{0_z}$.

The refracted pressure wave is measured by the piezotransducer installed exactly opposite the charge, near the wall of the rock at a distance of 3–5 mm from it. Such a gap may distort very slightly the shape of the stress wave due to its progression in water. The error may not exceed 1.5% [5].

Thus the total error of measurement does not exceed 10%, which is admissible if one considers that the total error in measurement including test conditions and the manufacturer's specifications of measuring instruments also lies within the limits.

From an analysis of assumptions and errors inherent in this method, it follows that while conducting experiments due attention should be paid to the accurate positioning of transducers exactly opposite the charge, perpendicular to the plane of refraction and at a distance not exceeding 5 mm from the wall of the block. Here, in the placement region of the transducer, the wall of area $S(\text{mm}^2) = 10\ b$ (where b is the width of the transducer) should ideally be smooth (polished). The remaining area (surface of wall) should be even, devoid of marked roughness and projections, particularly near the place of installation of the transducer. Otherwise the oscillogram being recorded will be distorted by various disturbances caused by diffraction.

Model experiments based on this method may be conducted in two ways: on the testing ground in conditions of natural deposition of rocks and in laboratory conditions on blocks made from rocks under investigation.

In the first method, under natural conditions, a monolithic section free from visible cracks and inhomogeneities and with a natural or artificial pit is selected. The vertical walls of the pit are smoothened and places around the installation of transducers are polished. The pit with installed transducers is filled with water. Charges, placed on lines perpendicular to the pit wall and at the installation points of transducers, are blasted.

Shot holes are drilled at varying distances from the pit wall. Charges are fired in series, starting from the farthest holes, so as not to damage the rock between the wall and the subsequent round. This method is rarely adopted as it requires a readily available pit with fairly even vertical walls. It is difficult to prepare an artificial pit without disturbing the continuity of the surrounding rock.

The *second method* is more suitable, i.e., laboratory experiments with blocks of the rock under consideration. Usually rectangular blocks with smooth sides are used. These are placed in a rectangular tank of water, roughly 2-3 times bigger than the blocks. Holes parallel to the sides of the block are drilled. The site of the hole is so chosen that during a blast measurements of the refracted wave may be conducted from four different distances using transducers installed on four different sides of the block.

It is essential to use tanks that are 2-3 times bigger than the dimensions of the blocks so as to exclude the effect of reflected waves from the tank walls on the refracted wave being recorded (Fig. 27).

Piezotransducers of pressure (usually tourmaline-based, as they have less pyro-effect) and special recording apparatus consisting of cathode ray oscillograph, block of cathode followers, control panel and feed blocks (Fig. 28) are used for measuring the parameters of a wave refracted in water. The principle of operation of this apparatus is as follows. Voltage (ac) induced by the pressing of plates in the transducer as the pressure wave passes, is fed to the block of cathode followers, intended for matching the wave impedance or signal voltage with the oscillograph. From the cathode

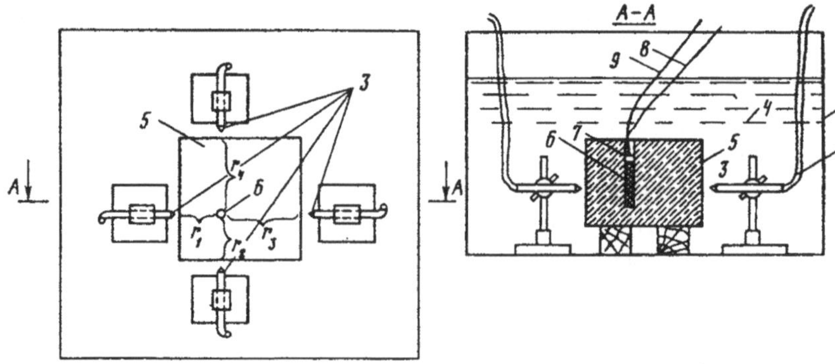

Fig. 27. Diagrammatic presentation of how to conduct laboratory experiments by the method of refracted wave.

1—tank; 2—measurement cable; 3—transducers; 4—water; 5—block of rock; 6—charge; 7—detonator; 8—lead wire; 9—firing cable.

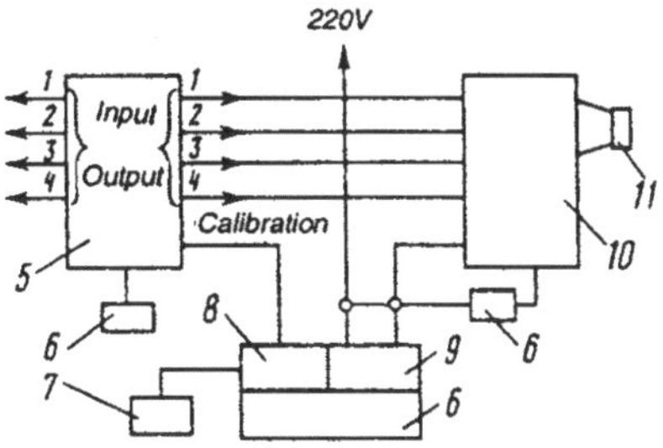

Fig. 28. Block diagram of measuring system.
1-4—piezotransducers of pressure; 5—block of cathode recorders; 6—power supply blocks; 7—millivoltmeter for measuring calibrated signal; 8—generator of calibrated pulses; 9—amplifier of triggering signals; 10—cathode-ray oscillograph; 11—camera.

followers, the inverted voltage is fed to the input terminal of the oscillograph through an armoured cable and fixed on the screen of the oscillograph by the cathode ray tube, in the form of an oscillogram of voltage variation over time. The recording thus obtained is deciphered as a diagram showing pressure variation over time, according to the formula

$$p = (A_{max}/A_K)(C_K U_K/S_{tr}), \qquad \ldots (3.34)$$

where p is the pressure in the refracted wave; A_{max} the maximum amplitude of the oscillogram; A_K amplitude of calibrated signal; C_K, U_K capacitance and voltage of calibrated signal; S_{tr} sensitivity of the piezotransducer.

The control panel consists mainly of two blocks: generator of calibrated pulses GKI-3 and amplifier of triggering signals (FUPS). These synchronise the activation of the oscillograph, emitting calibrated pulses and providing time markers for deciphering the oscillograms.

Time characteristics of recorded waves are deciphered by the formula

$$t = t_{sc} t^0,$$

where t is time; t_{sc} the time scale on the oscillogram; t^0 the time noted from the oscillogram.

As mentioned earlier, this method was developed for blasting concentrated charges to establish the radial component of a stress wave when the transducer was installed exactly opposite the charge, such that angles α in formula (3.33) were closer to zero. In the case of oblique incidence, the normal component given by relationship (3.33) would be recorded. If the ratio between σ_r and σ_φ, valid for the elastic zone of blast, is used in the form

$$\sigma_\varphi = (1 - 2r^2)\sigma_r = \frac{\nu}{1-\nu}\sigma_r,$$

where $\gamma = c_S/c_P$ the ratio between velocities of propagation of transverse and longitudinal waves and ν poisson ratio, then the radial component can be determined even for oblique incidence ($\alpha \neq 0$) from the relationships

$$\sigma_n = \sigma_r[\cos^2\alpha + (1 - 2\gamma^2)\sin^2\alpha] = p/k_{ref},$$

$$\sigma_{n.} = \sigma_r\left(\cos^2\alpha + \frac{\nu}{1-\nu}\sin^2\alpha\right) = p/k_{ref},$$

according to the formulae

$$\sigma_r = \frac{p}{k_{ref}} \cdot \frac{1}{1 - 2\gamma^2\sin^2\alpha}; \quad \frac{p}{k_{ref}} \cdot \frac{1-\nu}{\cos^2\alpha - \nu(2\cos^2\alpha - 1)} = \sigma_r.$$

But using the above method of refracted wave to determine the parameters of stress wave arising from firing an extended charge entrails taking into account certain factors that considerably complicate deciphering the experimental data and impose additional conditions for conducting the experiments.

First of all it should be noted that in a blast of extended charges initiated from one end, as actually done in practice, the shape of the stress wavefront would depend on the ratio of velocity of charge detonation D and velocity of propagation of longitudinal wave c_P.

If the detonation velocity is more than the propagation velocity of the longitudinal wave ($D > c_P$), the wavefront resembles the surface of a rotating pyriform body consisting of two semi-spheres that integrate with a cone. Angle α, formed between the generator of the cone (see, Fig. 8, a) and the charge axis, is determined from the ratio

$$\sin\alpha = c_P/D.$$

If $c_P > D$, then the wavefront is spherical in shape with its centre at the point of charge initiation (see, Fig. 8, b).

The normal-to-the-plane measurement of the stress wave component induced by firing an extended charge is determined by the method of refracted wave.

It should be borne in mind that while modelling on blocks, the dimensions have to be chosen properly, based not only on scale of geometric similitude but also by the condition of end effects of the block on the pattern of stress wave being recorded.

Knowing the approximate duration of the compressional phase of the wave from earlier experiments or publications, the dimensions of the block and location of the charge are so selected that waves reflected from the nearest face of the block pass by the observation point after passage of the compressional phase. The times of entry of direct and reflected waves are computed according to the laws of acoustics considering the geometry of mutual location of charge and observation point in the block and their distance from the sides and edges of the block.

It should be noted that the method of refracted wave allows one to determine fairly correctly only the positive phase of normal component of the stress wave, which is a distinct disadvantage. This is explained by the fact that water does not transmit rarefactions exceeding (1 to 2) 10^5 Pa and during refraction of the negative phase of the stress wave, cavitation may occur. Experimental oscillograms attest to this fact that after the compressional phase, usually chaotic, high-frequency vibrations are recorded near the zero line, which are characteristic of collapsing phenomena of the cavitational bulb.

To obtain the complete stress-time diagram, and also other components of the stress wave (σ_φ, σ_z) in the plane of measurement, other measurement methods need to be used, for example, direct measurement of stress wave parameters by placing piezotransducers of pressure in the rock specimen or in blocks made from equivalent materials of epoxy resin, acrylic plastic, sodium thiosulphate or sand-cement mixture. In this case the tranducers are placed in holes drilled in the blocks, affixing them firmly with fillers with acoustic stiffness nearly matching that of the material of the block. This is very easily done in blocks made from equivalent materials. Transducers are affixed in blocks made of acrylic plastic and resin using epoxy resin; in blocks made from sodium thiosulphate and cement-sand mixture the transducers are installed at the semi-liquid stage of the blocks prior to their cooling and hardening.

A *shortcoming* in this method of measurement is the possibility of obtaining only a qualitative picture of stress-wave diagram due to large difference in the physicomechanical properties, in particular the viscosity of the materials used and those of the rocks. Its *merit* lies in the possibility of obtaining complete stress-time diagrams of all the components of the stress wave by means of the corresponding positioning of sensitive planes of piezotransducers relative to the centre or axis of blast of spherical and cylindrical charges respectively. Thus in the blasting of a spherical charge, for recording the radial component of wave σ_r, the sensitive plane of the transducer is placed perpendicular to the direction of the blast centre, while for the tangential component $\sigma_\varphi = \sigma_\theta$ parallel to this direction.

The amplitude of stress wave is computed by formula (3.34), i.e., $\sigma_r, \varphi = (A_{max}/A_K)(c_K U_K/S_{tr})$, where r, are indices of radial and tangential components of the stress wave.

Apart from measuring stresses, several other methods are available for measuring the velocity of displacement or acceleration of rock particles in the stress wave. Capacitance-based, electromagnetic and induction converters are used for measuring the velocity of displacement. These are based respectively on the variation of voltage in the capacitor due to movement of the receiver plate relative to the immobile plate, on the appearance of a guided electric signal in the moving transducer (usually wire) in a uniform magnetic field and on the induced EMF (electromotive force) in the coil due to movement of the magnetic core within its gap.

Electromagnetic converters are widely used in laboratory conditions [10] and induction-type converters are widely used for measuring mass velocity and displacement of rock and soil in the far-field zone of a blast while recording blast-induced seismic waves.

Wire frames installed directly in the specimen and suitably oriented vis-à-vis the direction of the constant magnetic field are used as a transducer in the method of measurements with electromagnetic converters. Readily melting media (sodium thiosulphate, gum and different resins) are used as model material. The wire gauge can be affixed reliably in such materials prior to their cooling and hardening. In such cases the velocity of displacement of the medium (cm/s) is given by

$$v = U10^5/(lH),$$

where U is the voltage measured from the oscillograph; l the gauge length; H the magnetic field strength.

Using electromagnetic converters, it is possible to measure fairly accurately several components of particle velocity over a wide range of distances, from 3 to 70 times the charge radii [10]. The main disadvantages of this method are:

1) such converters cannot be used on rock specimens;
2) non-availability of direct calibration of the converters;
3) use of this method in full-scale conditions is not possible.

Accelerometers of different designs are used to measure rock particle accleration in the stress wave. More useful among those are the piezoquartz accelerometers of high strength and linearity of characteristics right up to or close to failure stress. These are high-frequency and unidirectional converters, i.e., electrical voltage generated by them is proportional to the acceleration in one direction—the axis of the converter. This is very important while conducting measurements in rocks, particularly in full-scale conditions, where rocks are essentially anisotropic media.

M.A. Nefedov developed piezoquartz accelerometers for measuring acceleration under laboratory and full-scale conditions [10]. Constructionally, the two are similar, differing only in weight and dimensions of the gauge, natural frequency, mechanical strength and sensitivity. Cathode ray oscillographs of the type OK-17, C1-33, C8-12 (usually in laboratory experiments) and mirror-galvanometer oscillographs of the type H-105, H-117 (for field tests and industrial experiments) are used as recorders. Cathode-ray oscillographs are very promising, with mechanical scanning of the image, for example, nine-channel oscillographs HO-23 and HO-40, being manufactured currently.

While using accelerometers, as in the case of pressure gauges, the problem of installation of gauges in the specimens or rock massif and their reliable fixture arises. The requirements for installation of accelerometers are similar to those imposed on piezo pressure gauges. To affix the transducers, an epoxy compound consisting of epoxy resin ED-20 (100 parts by weight), dibutylfluoride (12-15 parts by weight), finely grained quartz (80 parts by weight) and polyamino-polyethylene (12 parts by weight) are recommended [10]. This compound enables failure strength and acoustic stiffness very close to hard rocks ($\sigma_{rup} = 43$ MPa, $\rho c = 13 \times 10^6$ Pa/s).

The magnitude of acceleration is obtained from the relationship

$$a = (A_{max}/A_K)(C_K U_K/S_a),$$

where A_{max}, A_K, C_K, U_K are quantities similar to those in formula (3.34); a is the magnitude of acceleration expressed in units of acceleration due to gravity g; S_a the sensitivity of the accelerometer, as related to g.

It is to be noted that such measuring apparatus allow one to obtain the displacement velocity also, if in the measuring channel an integrater is introduced. This is very important for analysing one of the key parameters determining the blast effects on rock and engineering structures.

Graphic integration of recorded accelerations is quite complicated and lacks sufficient accuracy.

The conditions for modelling a stress wave while conducting experiments arise out of the extended law of geometric similitude based on the energetic similarity of *EC* and similarity of the modelling and full-scale media.

3.2.3 Methods of Modelling Air Blasts and Seismic Waves Induced by Large-scale Blasts in Rocks

In the far-field zone of blasting, the most important factors determining the blast effect on surface structures are air blast (AB) and blast-induced seismic wave (BSW). The latter affect underground structures also. Therefore, the parameters of such waves need to be studied. To determine these parameters (of AB and BSW), usually instrumental observations are carried out while conducting experimental blasts in industrial and field conditions.

Highly sensitive piezoelectric pressure gauges, compact power amplifiers in conjunction with cathode followers and mirror-galvanometer oscillograph H-117 with high-frequency galvanometers ($f_{gal} = 600$ to 1200 Hz) are used for measuring the parameters of *AB*. The selection of galvanometer is governed by the condition

$$f_{gal} \gg f_{AB} \approx 1/\tau_+,$$

where τ_+ is the duration of positive phase of *AB* established by the formula of M.P. Šadovsky,

$$\tau_+ = 1.5\sqrt[6]{G}\sqrt{r} \times 10^3.$$

Piezopressure gauges of knife or membrane types are used as transducers whose working frequency is equal to 100 and 10 kHz respectively. The power amplifier has a transmission band from 1 to 100 kHz. The entire set of apparatus (transducer-amplifier with cathode follower, mirror-galvanometer oscillograph) makes it possible to record *AB* over a wide range, from 10 to 10^6 Pa. The measured data, i.e., pressure amplitudes over time, is deciphered using formula (3.34). Small charges are used in investigations conducted in test polygons and the results of measurement are extrapolated to the full-scale using relationships arising out of the extended law of geometric similitude. The same law is used for conducting model experiments. In such a case the process of pressure variation over time is recorded by cathode-ray oscillographs C1-33, as the duration of processes in the model is considerably smaller than in the test-polygon and industrial blasts.

To determine the parameters of BSW, a series of instrumented observations are conducted along very typical and hazardous seismic profiles at three to five points, located at different distances from the blast site. Such parameters are recorded by a set of measuring apparatus, consisting of displacement, velocity or acceleration transducers for soil (rock) movement, measuring cable and oscillograph for fixing readings of transducers in the form of seismograms, i.e., recording the corresponding parameters of soil vibration over time. Induction converters and piezoaccelerometers are generally used as the transducers (seismic recorders). Vibration graphs of the type VIB-A, VEGIK manufactured by IFZ, Academy of Sciences, USSR, and differential velocity gauges of the type DDC, IDC and seismic receivers HC-3 are grouped into the first category. These are used in test blasts and field blasts. The piezoaccelerometers developed by the Leningrad Gornyi Institute (Mining Institute, Leningrad) may also be used and in laboratory conditions (transducers of smaller dimensions, $d = 15$ mm and $L = 20$ mm) the mirror-galvanometer oscillographs H-105, OSh-9, H-117 are used, in which the vibrations in electric signal induced by the transducers are recorded by the galvanometer and mirror-beam oscillograph system, on an optically sensitive paper moving at a particular speed. Tracing such vibrations over time thus becomes possible. In so doing, depending on the type of galvanometer—frequency of its natural vibrations f_{gal}, frequency of natural vibrations of the gauge f_{tr} and frequency of vibrations in soils f_{soil} being recorded—the measuring apparatus records the vibrations proportional to the displacement or rate of soil displacement over time. Thus for recording the velocity of displacement or magnitude of soil displacement, the following corresponding equations need to be fulfilled:

$$f_{gal} \gg f_{tr} < f_{soil}, \quad f_{gal} \geq f_{soil}.$$

Velocity of displacement or magnitude of displacement from the corresponding seismograms are calculated using the following relationships:

$$v = AR_{\Sigma}/(S_{tr}S_{gal}), u = AR_{\Sigma}/(S'_{tr}S_{gal}),$$

where v and u are amplitude of displacement velocity (cm/s) and movement of vibrations (cm) in soil, A the amplitude deflection on the seismogram, mm; R_{Σ} resistance in the measuring circuit (transducer, measuring cable, galvanometer), ohm; S_{gal} sensitivity of galvanometer, mm/mA; S_{tr}, S'_{tr} sensitivity of transducer, expressed in mV·s/cm and mV/cm respectively.

To measure accelerations in ground movement, usually large-size ($d = 40$ mm, $L = 45$ mm) accelerometers of the type LGI [10] are used. Readings of transducers are recorded on mirror-galvanometer oscillographs H-117 and the data processed according to formula (3.34). The modelling conditions for blast-induced seismic waves stem from the law of extended geometric similitude.

3.3 Modelling Shock and Stress Waves in Near-field Zone of a Blast

Development of explosion in rock in the near-field zone of the blast may be schematically represented as follows. When an explosive charge is initiated, a detonation with shock front is generated, followed by zone of reaction and field of expanding gaseous products of explosion. At the interface of *EC*-rock, due to the heavy compression of rock, a compressional wave with a shock front develops, progressing deeper into the rock symmetrically with reference to the charge. The energy of the compressional wave advancing in rock, rapidly attenuates and due to the nature of the relationship between deformations and stresses in rock, the compressional wave quickly loses the features of a shock wave and transforms initially into an elastoplastic, subsequently into an elastic wave with a non-shock front [5,6]. Gaseous products of explosion, while further expanding, compress the fragmented rock mass to form a crater cavity, if free faces are located at a fairly large distance from the charge, or a blast funnel if the explosion occurs near the free faces and the energy of gaseous products is adequate to overcome the forces of resistance offered by broken rock to the compressional wave as well as the quasi-static effect of explosion products.

Thus to model a shock wave and stress wave in the near-field zone of blast, it is necessary to establish the conditions of modelling the shock wave refracted at the charge rock interface, and the elastoplastic stress wave at certain distance from the charge. Obviously, the extended law of geometric similitude, valid for the zone of elastic deformations, will not be satisfied in the near-field zone due to the difference in boundary conditions, caused by the different types of *EC* and rock.

To model the refracted shock wave, let us consider the relationships derived in [4] from the conditions of refraction of detonation wave in rock, conditions of dynamic combination at the fronts of reflection and refraction waves and shock wave adiabatic expansion for hard rocks in the manner suggested by V.M. Gogolev. Maximum pressure in the refracted wave p_f according to [4] is determined from these equations:
for the case when acoustic stiffness of rock $\rho_p cr_p$ exceeds the impedance of $EC\ \rho_{EC} D$, i.e., $\rho_p c_p > \rho_{EC} D$,

$$\left\{\frac{p_f}{\rho_p}\left[1 - \frac{1}{\left(A\frac{p_f}{p_1 c_p^2} + 1\right)^{1/m}}\right]\right\}^{1/2} = v_1 - \frac{(p_f - p_1)\sqrt{2k}}{\sqrt{\rho_{EC}(k-1)[(k+1)p_f + (k-1)p_1]}}; \quad \ldots (3.35)$$

for the reverse case, i.e. $\rho_p c_p < \rho_{EC} D$,

$$\left\{\frac{p_f}{\rho_p}\left[1 - \frac{1}{\left(A\frac{p_f}{\rho_p c_p^2} + 1\right)^{1/m}}\right]\right\}^{1/2} = v_1 + \frac{2kD}{k^2 - 1}\left[1 - \left(\frac{p_f}{p_1}\right)^{\frac{k-1}{2k}}\right], \quad \ldots (3.36)$$

where p_1 is the maximum pressure in the detonation wave,

$$p_1 = \rho_{EC} D^2/(k+1);$$

A and m dimensionless numerical coefficients in the expression for the shock wave adiabatic process; $A = 5.5$; $m = 5$ for $0.1 \leqslant p_f/(\rho_p c_p^2) \leqslant 35, A = 3.0$; $m = 3$ for $p_f/(\rho_p c_p^2) < 0.1$; k the entropy index of detonation products of EC; v_1 the velocity of particles of explosion products in the detonation wave

$$v_1 = D/(k+1); \quad \ldots (3.37)$$

ρ_{EC} explosive charge density; D velocity of detonation of the explosive. The remaining parameters of the refracted wave, namely, density ρ_f, particle velocity v_f and velocity of front N_f, are determined by the formulae:

$$\rho_f = \rho_p\left(1 + \frac{Ap_f}{\rho_p c_p^2}\right)^{1/m}, \quad \ldots (3.38)$$

$$v_f^2 = p_f\left(\frac{1}{\rho_p} - \frac{1}{\rho_f}\right), \qquad \ldots (3.39)$$

$$N_f = \frac{p_f}{\rho_f v_f} = \sqrt{\frac{p_f}{\rho_f}\left(\frac{1}{\rho_f/(\rho_p - 1)}\right)}, \qquad \ldots (3.40)$$

From equations (3.35) and (3.36) it follows that for the model and the full-scale to be identical, it is necessary that the pressure scales $(p_f)_c$ and $(p_1)_c$ should be equal, $(p_f)_c = (p_1)_c = p_c$ and should satisfy the similitude criterion:

$$\frac{p_c}{(\rho_p)_c(c_p)_c^2} = 1. \qquad \ldots (3.41)$$

The scales of velocity at the detonation wave front $(v_1)_c$ and detonation velocity D_c and explosive density $(\rho_{EC})_c$ are related to the pressure scale p_c at $k_c = 1$ by the relationship

$$\frac{p_c}{(\rho_{EC})_c(v_1)_c^2} = 1. \qquad \ldots (3.42)$$

From relationship (3.37) it follows that $(v_1)_c = D_c = v_c$, i.e., the scales of velocity of explosion products at the detonation wavefront and velocity of detonation should be equal. In such a case the similitude criterion (3.42) is written in the form

$$\frac{p_c}{(\rho_{EC})_c D_c^2} = 1. \qquad \ldots (3.43)$$

From formulae (3.38) to (3.40), taking into consideration (3.41), it follows that

$$(v_f)_c = (N_f)_c = (c_p)_c = c_c, \qquad \ldots (3.44)$$

i.e., scales of modelling the velocity of rock particles at the shock front, velocity of propagation of the wave and sound in the rock should also be equal.

Comparing (3.41) and (3.43), the following criterial relationship is obtained

$$\rho_{EC} D^2/(\rho_{br}^2 c_{br}^2) = \text{ idem},$$

which relates the stiffness of rock $\rho_p c_p^2$ with a similar quantity $\rho_{EC} D^2$ of explosive charge and should be equal for both the model and the full-scale, i.e.,

$$\frac{(\rho_{EC})_N D_N^2}{(\rho_p)_N (c_p)_N^2} = \frac{(\rho_{EC})_M D_m^2}{(\rho_p)_m (c_p)_m^2}. \qquad \ldots (3.45)$$

From the similitude criteria (3.41), considering relationships (3.42) and (3.44), it follows that the parameters of movement of medium (rock or *EC*) in the refracted and detonation waves should satisfy the Euler similitude criterion $p/(\rho v^2) = \text{idem}$, which links the kinematic and dynamic characteristics of such movement. If in such a case simple geometric similitude is valid, then the scales of linear dimensions l_c and time t_c should be equal, i.e., $l_c = t_c$, and a block of rock is used as the model of the medium, i.e.,

$$(\rho_p)_N = (\rho_p)_M; \ (c_p)_N = (c_p)_M,$$

then equality (3.45) reduces to the following

$$(\rho_{EC})_M D_m^2 = (\rho_{EC})_N D_N^2. \qquad \ldots (3.46)$$

This equality is satisfied if in the model an explosive similar in charge density to that in full-scale blasts is used. However, using industrial explosives under laboratory conditions is not feasible, as the very small dimensions of charges in the model would be lower than the critical diameter of charges of such explosives, at which detonation would be stable and dependable. Usually, in laboratory model experiments small charges of other explosives (d_z = 3–10 mm) are used, such as TEN and Hexagene, which can be reliably detonated. In this case, knowing the parameters of the industrial explosive being used in the model, according to relationship (3.46) the density and detonation velocity of the model *EC* are chosen; in other words, the type of model *EC* is selected. When TEN and Hexagene are used in granular form, it is necessary to reduce the charge density of the model charges by

$$(\rho_{EC})_M = (\rho_{EC})_N (D_N/D_M)^2.$$

Thus, if $D_N = 4000$ m/s, $(\rho_{EC})_H = 1$ g/cm^3, $D_M = 6000$ m/s, then $(\rho_{EC})_M = 1(4/6)^2 \approx 0.45$ g/cm^3. But at such low density, the possibility of reduction in detonation velocity of the model charge exists, which may be approximately evaluated by the Cooke relation,

$$D = D_1 + 3500(\rho_{EC} - 1), \qquad \ldots (3.47)$$

where D is detonation velocity of *EC* at $\rho_{EC} = 1$ g/cm^3. In this case the charge density is computed using the following equation

$$(\rho_{EC})_M^3 + 2\left(\frac{D_M}{3500} - 1\right)(\rho_{EC})_M^2 + \left(\frac{D_M}{3500} - 1\right)(\rho_{EC})_M \times \frac{(\rho_{EC})_N D_N^2}{(3500)^2} = 0. \qquad \ldots (3.48)$$

Here $(\rho_{EC})_M = 0.66$ g/cm^3.

The parameters of the refracted wavefront may be modelled on rock specimens of spherical or cylindrical shape, and granular TEN or Hexagene used as the *EC* with density and detonation velocity calculated according to formulae (3.47) and (3.48). Model experiments should preferably be conducted on an X-ray pulse device (see Sec. 3.2.1) with recording of velocity of rock particle movement at the wavefront v_f. Here the movement of column wall at the initial stage of its development is recorded with minimum possible delay in activating X-ray tubes from the moment of explosion. Determining displacement of the column wall in the chosen section from the X-ray photofilms $\Delta S_c = (d_1 - d_{in})/2$ (where d_{in} is initial diameter of the column, d_1 its subsequent diameter after the wall has moved) and knowing the time interval between exposures Δt, the mean velocity of movement of column at the initial moment of explosion $v_{mc} = \Delta S_c/\Delta t$ is computed. Later, the rate of displacement of external surface of the test specimen at the time of passage of the shock wave, is determined in a similar manner. Considering that the shock wave in spalled rocks exists at a distance of up to 2 to 5 times the charge radii, the diameter of rock specimen is taken as equal to 3 to 6 times the charge radii, i.e., $d_{ts} = (3 \text{ to } 6)R_{0_z}$. The average displacement of specimen surface is given by

$$\Delta S_K = \frac{1}{2}[(d_{ts})_1 - (d_{ts})_{in}],$$

(where the indices '*in*' and '1' denote the value of d_{ts} before the blast and after passage of the shock wave on the surface of the specimen). Velocity of displacement of the edge $v_{ed} = \Delta S_K/\Delta t$ and velocity at the front of the wave on the specimen edge under consideration

$$v_f = v_{ed}/2 = \Delta S_K/(2\Delta t).$$

For accurate determination of the initial velocity of rock particles in the shock wave $(v_f)_0$, experiments are conducted on specimens of at least three sizes, $d_{ts} = 3R_{0_z}$, $4R_{0_z}$ and $6R_{0_z}$. From the experimental results four values of v_f are obtained [for different values of d_{ts}—Translator]:

$$v_f = v_{mc};\ v_f = v_{ed}/2 (\text{at } d_{ts} = 3 \text{ to } 6R_{0_z}).$$

Based on these, a curve showing the variation in v_f with distance is plotted. By measuring the value $v_f = v_{mc}$ at a distance, $r = R_{0_z} + \Delta S_c/2$, $v_f = v_{ed}/2$ at distances $r = (3 \text{ to } 6)R_{0_z}$ respectively. The value of $(v_f)_0$ in the refracted wave is obtained from this curve by extrapolating it for $r = R_{0_z}$, and the pattern of damping of amplitude of the shock front is determined.

The remaining parameters of the shock wave, p_f, ρ_f, N_f, are computed based on relationships (3.38) to (3.40), assuming the values of A and m, from the approximate magnitude of pressure in the refracted wave, according to the formula,

$$p_f = p_1 \frac{2\rho_p c_p}{\rho_p c_p + \rho_{EC} D}.$$

Geometrical scale of modelling l_c is taken from the diameter of charges in the model and the full-scale $(R^*_{0_z})_M$ and $(R_{0_z})_N$

$$l_c = \frac{(R_{0_z})_N}{(R^*_{0_z})_M} = \frac{(R_{0_z})_N}{(R_{0_z})_M}\left[\frac{(\rho_{EC})_N Q_N}{(\rho_{EC})_M Q_M}\right]^{1/2},$$

where $(R_{0_z})_M$ is the actual radius of model charge; $(R^*_{0_z})_M$ equivalent radius of model charge, determined from formulae (2.28) at $n = 2$ or (2.29) at $n = 3$.

The stress wave in the near-field zone of blast ($\bar{r} \leqslant 10$) is associated with elastoplastic deformations and large dissipative losses typical of the phenomena occurring under heavy pressures and high temperatures. For modelling a stress wave in this range of distances from the charge, where elastic and plastic deformations occur, the elements of viscosity and plasticity of the media being modelled must be similar. In the case of elastoviscous medium for a plane wave, equation of motion in the direction of propagation (axis x) is of the following form

$$\rho\frac{\partial^2 u}{\partial t^2} = (\lambda + 2\mu)\frac{\partial^2 u}{\partial x^2} + \frac{4}{3}\eta\frac{\partial^3 u}{\partial x^2 \partial t}, \qquad \ldots (3.49)$$

where η is the coefficient of dynamic viscosity of the medium; u the displacement of rock particles, ρ the density of rock; λ, μ—Lamé's constants.

This equation, in contrast to the equation for an elastic medium, has an additional viscosity term on the right side giving rise to a new similitude criterion, Reynolds number. Actually, if similar transformations are done for equation (3.49), we obtain identically equal complexes, consisting of similitude scales in the form

$$\rho_c l_c / t_c^2 = \lambda_c / l_c = \mu_c / l_c = \eta_c / (l_c t_c). \qquad \ldots (3.50)$$

From this system of equalities (3.50), a series of similitude indicators may be obtained. If the second or third term is divided by the first, we get

$$\lambda_c/(\rho_c l_c^2/t_c^2) = \mu_c/(\rho_c l_c^2/t_c^2) = \lambda_c/(\rho_c v_c^2) = \mu_c/(\rho_c v_c^2) = 1,$$

which gives the Euler criterion $\sigma/(\rho v^2) =$ idem. As Lamé's constants λ and μ characterise the elastic properties of rock, these are expressed through the modulus of elasticity E and Poisson ratio ν:

$$\lambda = E\nu/[(1+\nu)(1-2\nu)], \; \mu = E\nu/[2(1+\nu)].$$

From the first and last terms of these equalities, we get a similitude indicator of the type

$$\frac{\rho_c l_c}{t_c^2} \div \frac{\eta_c}{l_c t_c} = \frac{\rho_c l_c^2}{\eta_c t_c} = \frac{v_c l_c}{\eta_c/p_c} = \frac{v_c l_c}{\zeta_c} = 1,$$

which is the similitude indicator of the Reynolds number $Re = vl/\zeta$, where ζ is the coefficient of kinematic viscosity. The scale for coefficient of kinematic viscosity is established from the relationship,

$$\zeta_c = v_c l_c = c_c l_c, \qquad \ldots (3.51)$$

We can obtain the same result by considering elastoplastic media or elastic media with residual deformations. The equation of motion in such a medium for a plane wave is described by the expression

$$\rho \frac{\partial^2 u}{\partial t^2} = (\lambda + 2\mu) \frac{\partial^2 u}{\partial x^2} + \frac{(\lambda + 2\mu)}{4/3\eta} \frac{\partial^3 u}{\partial x^2 \partial t}, \qquad \ldots (3.52)$$

from which by an analogous method of transformations, the similitude indicator of the Reynolds number and expression (3.51) for the scale of the coefficient of kinematic viscosity can be obtained. Here the scale of the coefficient of dynamic viscosity would be directly proportional to the scale of acoustic stiffness of the media $\rho_c c_c$ and to the scale of geometric similitude, i.e.,

$$\eta_c = \rho_c c_c l_c. \qquad \ldots (3.53)$$

The Euler criterion, as shown in [5, 6] allows reducing the conditions of modelling to those arising out of the extended law of geometric similitude, when the time scale and linear scale are equal, i.e., $l_c = t_c$, and a specimen of the rock under investigation is used as the model specimen ($\rho_c = 1$). In such a case the scales of velocity v_c and stresses σ_c are also equal to unity. However, to satisfy the Reynolds similitude criterion conditions (3.51) or (3.53) must be met. Obviously, while modelling viscoelastic motion, such should be conducted in a less viscous medium than the full-scale medium. In such a case, modelling in the full-scale medium, strictly speaking, is wrought with large scale-effects or errors. Such modelling is possible only for brittle spalled rocks in which viscosity is almost non-existent and the last term in equations (3.49) and (3.52) can be neglected. For semi-spalled rocks and highly viscous rocks, modelling should be done on equivalent material specimens whose physicomechanical properties are chosen from combinations (3.51) or (3.53), which in an expanded form are described by equalities:

$$\zeta_M = \zeta_N \frac{(c_p)_M}{(c_p)_N l_c}, \qquad \ldots (3.54)$$

$$\eta_M = \eta_N \frac{(\rho_p)_M (c_p)_M}{(\rho_p)_N (c_p)_N l_c}. \qquad \ldots (3.55)$$

Thus for modelling a stress wave in the near-field zone of a blast ($\overline{n} \leqslant 10$), i.e., in the zone of elastoplastic deformations, for semi-spalled rocks and highly viscous rocks it is necessary to use equivalent materials, identified from condition (3.54) or (3.55), under arbitrarily assigned scales of geometric similitude l_c, density ρ_c or velocity c_c. Here the method of either refracted wave, or the induction method or the method of measurement using accelerometers, as well as other methods of studying the processes of rock breakage in the near-field zone of blasting (X-ray graphics or optical dynamics) may be adopted.

3.4 Modelling Development of Crater and Rupture Zones in Camouflet Blasting

It is well known that in the formation and development of a crater cavity a key role is played by gaseous products of the blast, while in rock fragmentation and formation of breakage zones both compressional stress waves and quasi-static pressure of explosion products take part.

Let us start with initiation and formation of a crater. The solution to this problem, as suggested by E.I. Shemyakin, is used for a blast of spherical charge in a model of non-compressible elastoplastic medium, wherein the cavity expands due to shear deformations while triaxial deformations are non-existent. Here the dilatations associated with the formation of fractures and effect of heat exchange are not taken into account, assuming the expansion to be an adiabatic process occurring under the effect of internal pressure p_0 of blast products, and varying as per the law

$$p_0 = p_{00}(R_{0_z}/R_0)^{3n},$$

where p_{00} is the conditional initial average pressure in the charge column (for a standard charge of Trotyl $\rho_{EC} = 1.6$ g/cm^3, $p_{00} = 8.5 \times 10^8$ Pa); R_{0_z} the radius of charge (charge column); R_0 current radius of the cavity; n an exponent of the adiabatic expansion of blast products ($n = 1.25$ for the asymptotic branch of the adiabatic expansion chosen in the given problem).

The deformed zone around the charge is divided into two zones (plastic and elastic), demarcated by a conditional sphere of radius R_*. Ground movement is characterised by the velocity of rock particle displacement $v(r, t)$ while the deformations are characterised by shear deformations γ, and the stress state by principal stresses σ_r and $\sigma_\varphi = \sigma_\theta$ which, in spherical co-ordinates with origin at the centre of charge (cavity), happen to be the corresponding radial and tangential components of the stress wave.

The principal shear deformations γ are related to principal tangential stress τ by Hooke's law

$$\tau = \mu\gamma, \quad \mu = \rho c_S^2,$$

where μ is the shear modulus; c_S velocity of propagation of transverse waves. From this the ultimate value of principal shear deformation is given by

$$\gamma_e = \tau_s/\mu,$$

where τ_s is the ultimate yield strength under shear.

Within the zone of elastic deformations, the process of stress relief obeys Hooke's law, i.e.,

$$\tau = \mu\gamma, \tau \leqslant \tau_s \text{ and } \gamma = \gamma_e.$$

In the zone of plastic deformations this is determined by the relationship

$$\tau_s - \tau = \mu(\gamma_e - \gamma).$$

The plasticity condition is the ideal condition for yielding

$$\sigma_\varphi - \sigma_n = 2\tau_s.$$

For a specific solution to the problem, the following boundary and initial conditions are assigned.
At the boundary of expanding cavity at $r = R_0(t)$

$$v_p(R_0) = \frac{dR_0}{dt}; \; -\sigma_r(R_0) = p_0.$$

In the sphere, demarcating the plastic and elastic zones, at $r = R_*(t)$

$$-p = \sigma_e + \frac{4}{3}\tau_s; \; v_p(R_*) = v_e(R_*)$$

As $r \to \infty, p \to p_h$ (where $p_h = \rho g H$, the rock pressure).

At the initial movement of cavity, it is assumed that a certain initial conditional velocity v_0 is imparted to the cavity wall, i.e., at $t = 0$ and $R_0 = R_{0_z}$, $dR_0/dt = v_0$. With such a formulation of the problem, after an entire series of transformations, E.I. Shemyakin obtained the final relationship for the mechanism of movement of radius of cavity over time $R_0(t)$, which may be represented in the following manner for easy determination of similitude criteria:

$$dR_0/dt = \left\{\left\{v_0^2\left(\frac{R_{0_z}}{R_0}\right)^3 + \frac{2}{3} \times \frac{1}{\rho}\left(\frac{R_{0_z}}{R_0}\right)^3 \times \right.\right.$$
$$\times \left\{\frac{p_{00}}{n-1} \times \left[1 - \left(\frac{R_{0_z}}{R_0}\right)^{3(n-1)}\right] + p_h\left[1 - \left(\frac{R_0}{R_{0_z}}\right)^3\right]\right\}$$
$$\left.\left. - \frac{8}{9} \times \frac{\tau_s}{\rho}\left[1 - \left(\frac{R_{0_z}}{R_0}\right)^3\right] \times \left(\ln\frac{1}{\gamma_e} + 1\right)\right]\right\}\right\}^{1/2}. \quad \ldots (3.56)$$

If the method of similar transformations is adopted and (3.56) is rewritten for the model, considering the similitude scales of quantities contained in it, we get

$$\frac{l_c}{t_c}\left(\frac{dR_0}{dt}\right)_M = \left\{v_c^2 v_{0M}^2\left(\frac{R_{0_z}}{R_0}\right)_M + \frac{2}{3\rho_c}\frac{1}{\rho_M}\left(\frac{R_{0_z}}{R_0}\right)_M^3 \times \left\{\frac{(p_{00})_c(p_{00})_M}{n-1} \times \right.\right.$$

$$\times\left[1-\left(\frac{R_{0_z}}{R_0}\right)_M^{3(n-1)}\right] + (p_h)_c(p_h)_M = 1 - \left(\frac{R_0}{R_{0_z}}\right)_M^3\Bigg\}$$

$$\left.\left. -\frac{8}{9}\frac{(\tau_s)_c(\tau_s)_M}{\rho_c\rho_M} \times \left[1-\left(\frac{R_{0_z}}{R_0}\right)_M^3\right]\left(\ln\frac{1}{\gamma_e}+1\right)\right\}\right\}^{1/2} \quad \ldots (3.57)$$

Whence for expressions (3.56) and (3.57) to be identical, it is necessary to equate the scale complexes of all terms right and left in (3.57). So, we have

$$l_c^2/t_c^2 = v_c^2 = (p_{00})_c/\rho_c = (\tau_s)_c/\rho_c = (p_h)_c/\rho_c = q_c h_c. \quad \ldots (3.58)$$

Now it is easy to obtain the similitude indicators by relating successively the first, third and fourth terms of equality (3.58) to the second. This gives

$$v_c t_c/l_c = 1,\ (p_{00})_c/(\rho_c v_c^2) = 1,\ (\tau_s)_c/(\rho_c v_c^2) = 1,$$
$$g_c h_c/(v_c^2) = g_c l_c/v_c^2 = 1. \quad \ldots (3.59)$$

Under conditions of similar media in the model and the full-scale and fulfilling the geometric similitude for linear dimensions and time, i.e., $l_c = t_c$, from (3.59) we get

$$\rho_c = 1, v_c = 1,\ (\tau_c)_c = 1, (p_{00})_c = 1,$$
$$g_c h_c = 1 \text{ or } g_N h_N = g_M h_M. \quad \ldots (3.60)$$

Thus it follows that when the law of extended geometric similitude is satisfied and the same media are used in the model and the full-scale and the necessary condition $(p_{00})_c = 1$ for selection of *EC* as well as condition (3.60) requiring weightage of the model according to the relationship $g_M = g_N l_c$; $(\gamma_M = \gamma_N l_c)$ (for example, in the centrifugal modelling method) are satisfied, then the cavity's movement would be proportional to the scale of linear dimensions $((R_0)_N = (R_0)_M l_c)$, and the rate of cavity movement would be equal in both, i.e., $(dR_0/dt)_N = (dR_0/dt)_M$.

In the case of blasting at shallow depths and the condition $\tau_s \gg p_h$ being satisfied (which is usually fulfilled for spalled rocks), the last condition in 3.60 $(g_N h_N = g_M h_M)$ may be neglected. It then becomes possible to model the cavity development using rock specimens by taking into account the conditions of extended law of geometric similitude and selection of *EC* from the condition $(p_{00})_M = (p_{00})_N$, resulting in the fulfilment of condition (3.47) and solution of equation (3.48).

Thus it is preferable to model the development of a crater cavity using rock specimens by the X-ray pulse device, according to the method described in Sec. 3.2.1. In this case, to model a blast at a great depth, wherein rock pressure exerts a significant effect (p_h is commensurate with τ_s), it is necessary to load the blast column by static pressure of gas $p_{stat} = (p_h)_N$ (blasting to be conducted in a compression chamber). In both cases, while choosing the linear scale l_c and charge dimensions, one has to start from a maximum possible thickness of rock specimen $(d_{ts})_{max}$ which still allows penetration of X-rays, and also from the condition ensuring absence of the effect of stress wave reflected from the surface of the specimen. Regarding the latter, the time of entry of the reflected stress wave at the wall of the cavity that has completely taken shape t_{ref}, should be more than the time taken for development of the cavity t_{cav}, i.e., $t_{ref} > t_{cav}$.

The time of entry of reflected stress wave

$$t_{ref} \approx (d_{ts} - R_{0_z} - R_{0_m})/c_p. \qquad \ldots (3.61)$$

The duration over which the cavity develops to its maximum size R_{0_m} is equal to

$$t_{cav} = (R_{0_m} - R_{0_z})/(v_{cav})_{mc} \approx (R_{0_m} - R_{0_z})/[0.1 \text{ to } 0.2)c_p], \qquad \ldots (3.62)$$

where $(v_{cav})_{mc}$ is the average rate of cavity development, equal to 0.1 c_p for spalled rocks and 0.2 c_p for unspalled rocks [9].

From formulae (3.61) and (3.62), we have for spalled rocks $(R_{0_m} \leqslant 1.5R_{0_z})$

$$(d_{ts})_{max} \geq 7.5R_{0_z}, R_{0_z} \leqslant 0.13(d_{ts})_{max}$$

for unspalled rocks $(R_{0_m} \leqslant 3R_{0_z})$

$$(d_{ts})_{max} \geq 14R_{0_z}, R_{0_z} \leqslant 0.07(d_{ts})_{max}.$$

Let us consider the modelling of crushed zones in a camouflet blast. By crushed zones we mean both the zone of crushing or pulverisation of rock and the fractured zone, characterised mainly by the zone of radial cracks.

The *crushed zone or pulverised zone* is directly adjacent to the charge column and extends to a distance of (5 to 10) R_{0_z}. It is induced mainly by the action of shock wave and a powerful elastoplastic wave of compression. Hence the conditions for modelling this zone are those imposed on modelling the shock and stress waves as derived in Sec. 3.3.

Let us examine in detail the conditions for modelling the *fractured zone,* formed due to the action of stress wave and quasi-static pressure exerted by products of explosion. We shall consider this for a blast of spherical (concentrated) charge.

The failure zone of the massif (fractured zone), formed by action of the stress wave, is determined by the effect of the rarefaction phase of the tangential component σ_{φ} of the wave, that widens and initiates radial cracks afresh if the following condition is satisfied:

$$\sigma_{\varphi max} = \left[1 - 2(c_S/c_P)^2\right]\sigma_{r max} = \frac{\nu}{1-\nu}\sigma_{r max} \geq \left[\sigma_{dn rup}\right], \qquad \ldots (3.63)$$

where c_P and c_S are propagation velocity of longitudinal and transverse waves respectively; ν Poisson ratio ($\nu = \varepsilon_x/\varepsilon_y$); $\sigma_{dn rup}$ dynamic tensile strength of rock.

To model the stress wave under the same media in the full-scale and the model, as already shown in Sec. 2, it suffices if the law of extended geometric similitude (for different *EC* used in the full-scale and the model) or simply geometric similitude (when the same *EC* is used in the full-scale and model) is fulfilled. In such a case $l_c = t_c$ is selected either from the condition of charge dimensions in the model, i.e., $l_c = (R_{0_z})_N/(R^*_{0_z})_M$, or from the typical dimensions of the model, for example, dimensions of rock block in the model. Then the scale of stresses of interest to us here would be $\sigma_N/(\rho_N v_N^2) = \sigma_M(\rho_M v_M^2)$ as $v_c = 1$ at $l_c = t_c(v_c = l_c/t_c = 1)$, under the same media ($\rho_c = 1$) and the Euler criterion $\sigma/\rho v^2 =$ idem. From this, we have

$$\sigma_c = \sigma_N/\sigma_M = \rho_N/\rho_M(v_N/v_M)^2 = 1.$$

However, in this case, to avoid the effect of stress waves reflected from the sides of the block reaching the fractured zone, it is essential that the zone of chipping b_{ch} from the side of the block not reach the fractured zone R_{frac} starting from the centre of the charge (axis), i.e., the following condition for block length L is to be maintained:

$$L/2 \geq R_{frac} + b_{ch} \text{ or } b_{ch} \leqslant L/2 - R_{frac}. \qquad \ldots (3.64)$$

For concentrated and spherical charges [6], taking into account expression (3.63), we have

$$\sigma_{\varphi max} = \frac{\nu}{1-\nu}\rho_p c_p \frac{2.25\times 10^3}{\overline{R}^2_{frac}} \geq \left[\sigma_{dn rup}\right],$$

hence, $$\overline{R}_{frac} \leqslant \left(\frac{\nu}{1-\nu}\rho_p c_p \frac{2.25\times 10^3}{\left[\sigma_{dn rup}\right]}\right)^{0.5}, \; R_{frac} = \overline{R}_{frac} \times R^*_{0_z}.$$

The zone of chipping is determined from the condition that the reflected radial component of the stress wave exceed the dynamic ultimate tensile strength, i.e., from the condition

$$\sigma_{r max}(R_{ch}) = \sigma_{r max}\left(\frac{L}{2} + b_{ch}\right) = \frac{2.25\times 10^3 \rho_p c_p}{\left(\frac{\overline{L}}{2} + \overline{b}_{ch}\right)^2} \geq \left[\sigma_{dn rup}\right]. \qquad \ldots (3.65)$$

Considering condition (3.64) and assuming that to satisfy it, meeting the following equality suffices, $L/2 = R_{frac} + b_{ch} + 10R_{0_z}$ or $b_{ch} = L/2 - 10R_{0_z} - R_{frac}$, formula (3.65) is written in the form

$$\sigma_{r_{max}}(\overline{L} - \overline{R}_{frac} - 10) = \frac{2.25 \times 10^3 \rho_p c_p}{(\overline{L} - \overline{R}_{frac} - 10)^2} = \left[\sigma_{dn_{rup}}\right],$$

hence

$$\overline{L} = \left[\left(\rho_p c_p \frac{2.25 \times 10^3}{\left[\sigma_{dn_{rup}}\right]}\right)^{0.5} + \overline{R}_{frac} + 10\right].$$

Let us analyse the modelling of a crushed zone induced by quasi-static pressure of gaseous products of an explosion by a simplified idealised scheme of blast effect on rock. According to this scheme, the rock is assumed to be incompressible while deformations in it are considered to be unit displacements.

Let there be a spherical camouflet cavity of radius R_0 as a result of an explosion (R_{0_z}—initial radius of charge column), and let individual elements of the rock be displaced in radial directions by a quantity S_R at a distance R from the blast centre.

Then for incompressible rock we have

$$\frac{4}{3}\pi R_0^3 - \frac{4}{3}\pi R_{0_z}^3 = \frac{4}{3}\pi (R + S_R)^3 - \frac{4}{3}\pi R^3,$$

hence

$$\left(\frac{R_0}{R_{0_z}}\right)^3 - 1 = \left(\frac{R}{R_{0_z}} + \frac{S_R}{R_{0_z}}\right)^3 - \left(\frac{R}{R_0}\right)^3. \qquad \ldots (3.66)$$

Here the energy transmitted to rock in a blast through the radius of camouflet cavity would be

$$\frac{4}{3}\pi \left(R_0^3 - R_{0_z}^3\right)(\sigma_* + \rho g H + p_{atm}) = \frac{p_0 V}{\kappa - 1}, \qquad \ldots (3.67)$$

where σ_* the stress value characterising rock strength; H the height of rock column above the blast site; ρ and g the rock density and acceleration due to gravity; p_{atm} the atmospheric pressure; p_0 the quasi-static pressure of blast products within the cavity created, $p_0 = p_{00}(R_{0_z}/R_0)^{3\kappa}$; V the volume of crater cavity, $V = 4\pi R_0^3/3$; κ the index of adiabatic expansion of blast products, equal to 1.25 for condensed EC.

After substituting the expressions for p_0 and V in the right side, equation (3.67), gets transformed to

$$\frac{p_0 V}{\kappa - 1} = \frac{4}{3}\pi \frac{p_{00}}{\kappa - 1}\left(\frac{R_{0_z}}{R_0}\right)^{3\kappa} R_0 = \frac{1\pi p_{00}}{3\kappa - 1}\frac{R_{0_z}^{3\kappa}}{R_0^{3(\kappa-1)}},$$

or taking $\kappa = 1.25$

$$\frac{p_0 V}{\kappa - 1} = \frac{4}{3}\pi \frac{p_{00} R_{0_z}^{3.75}}{0.25 R_0^{0.75}} = \frac{16}{3}\pi p_{00} \frac{R_{0_z}^3}{\overline{R}_0^{0.75}}.$$

Substituting this in equation (3.67), we get

$$\left(\frac{R_0^3}{R_{0_z}^3} - 1\right)(\sigma_* + \rho g H + p_{atm}) = \frac{4 p_{00}}{\overline{R}_0^{0.75}}. \qquad \ldots (3.68)$$

Expanding equality (3.66), we have

$$\left(\frac{R_0}{R_{0_z}}\right)^3 - 1 = 3\left(\frac{R}{R_{0_z}}\right)^2 \frac{S_R}{S_{0_z}} + 3\left(\frac{R}{R_{0_z}}\right)\left(\frac{S_R}{R_{0_z}}\right)^2 + \left(\frac{S_R}{R_{0_z}}\right)^3.$$

As $\dfrac{S_R}{R_{0_z}} \ll 1$, we can write the above equation as

$$\left(\frac{R_0}{R_{0_z}}\right)^3 - 1 = 3\,(R/R_{0_z})^2\,(S_R/R_0). \qquad \ldots (3.69)$$

Substituting (3.69) in (3.68), we get

$$\left(\frac{R}{R_{0_z}}\right)^2 \times \frac{S_R}{R_{0_z}} = \frac{4 p_{00} \overline{R}_0^{-0.75}}{3(\sigma_* + \rho g H + p_{atm})}.$$

Hence,
$$S_R = \frac{4 p_{00} R_{0_z}^3 \overline{R}_0^{-0.75}}{3(\sigma_* + \rho g H + p_0) R^2}.$$

Let us assume that rock gets fragmented at $\delta = S_R/R \geq \delta_{ed}$. Then, substituting $S_R = \delta_{ed}$ and dividing both sides by R, we get

$$\delta_{ed} = \frac{4 p_{00} \overline{R}_0^{-0.75} R_{0_z}^3}{3(\sigma_* + \rho g H + p_{atm}) R^3} = \frac{4}{3} \frac{p_{00} \overline{R}_0^{-0.75}}{(\sigma_* + \rho g H + p_{atm}) \overline{R}^3}$$

or
$$\overline{R}_p^3 = \frac{4}{3} \frac{p_{00} \overline{R}_0^{-0.75}}{(\sigma_* + \rho g H + p_{atm}) \delta_{ed}}$$

$$\overline{R}_p = \left[\frac{4}{3} \frac{p_{00} \overline{R}_0^{-0.75}}{(\sigma_* + \rho g H + p_{atm}) \delta_{ed}}\right]^{1/3}; \; R_p = \overline{R}_p R_{0_z}. \qquad \ldots (3.70)$$

Under one and the same medium, with geometric similitude being maintained,

$$\rho_N = \rho_M; \; \sigma_N^* = \sigma_M^*; \; p_{atm_N} = p_{atm_M} \text{ and } (\delta_{ed})_N = (\delta_{ed})_M$$

and for modelling, the essential conditions are

$$\left(p_{00}\overline{R}_0^{-0.75}\right)_N = (p_{00}\overline{R}_0^{-0.75})_M, \qquad \ldots (3.71)$$

$$(gH)_N = (gH)_M. \qquad \ldots (3.72)$$

In modelling with a specimen from natural rock and taking into account the law of extended geometric similitude, condition (3.71) is satisfied under the condition $(p_{00})_N = (p_{00})_M$, as according to the data published in [4], R_{0_M} may be written as

$$\overline{R}_{0_M} = \frac{1310}{(\rho c_S^2)^{0.292}}, \qquad \ldots (3.73)$$

from which it is seen that $(\overline{R}_0)_N = (\overline{R}_0)_M$ when the media for the model and the full-scale are similar $(\rho_N = \rho_M; c_{S_N} = c_{S_M})$.

The condition $(p_{00})_N = (p_{00})_M$ leads to the selection of *EC* according to condition (3.47) and to the solution of cubic equation (3.48).

Let us consider fulfilment of condition (3.72). We have

(a) $g_M = g_N \frac{H_N}{H_M} = g_N l_c$, i.e., condition (3.72) led to the condition of centrifugal modelling;

(b) if $\sigma_* \gg \rho g H$ (neglecting rock pressure), i.e., condition (3.72) may be neglected, then the second equation (3.70) may be written in the form

$$\overline{R}_p = \left[\frac{4}{3}\frac{p_{00}\overline{R}^{-0.75}}{(\sigma_* + p_{atm})\delta_{ed}}\right]^{1/3}.$$

In such a case modelling can be done using the law of extended geometric similitude and condition (3.46).

In both cases dimensions of the model should be such that the zone of chipping phenomenon does not extend to the crushed zone, i.e., condition (3.64) should be satisfied which, in this case, would be

$$L/2 \geq R_p + b_{ch}; \; b_{ch} \leqslant L/2 - R_p.$$

For rocks, usually $\overline{R}_p \leqslant 30\text{–}40$. If $\overline{R}_p = 40$ to 50, taking into account condition (3.66) and equating $\sigma_{r_{max}}^{rup}$ to ultimate tensile strength $[\sigma_{dn_{rup}}]$, we have

$$\sigma_{r_{max}}^{rup}(L/2 + b_{ch}) = \sigma_{r_{max}}^{rup}\left(L - R_p\right) = \left[\sigma_{dn_{rup}}\right];$$

now it is possible to determine the length of rock block L from the relationships describing $\sigma_{r_{max}}^{rup}(L - R_p)$.

For a spherical charge

$$\sigma_{r_{max}}^{rup} = \rho_p c_p \frac{2.25 \times 10^3}{(\overline{L} - \overline{R}_p)^2} = \rho_p c_p \frac{2.25 \times 10^3}{(\overline{L} - 40)^2} = \left[\sigma_{dn_{rup}}\right],$$

hence, $$\overline{L} = 40 + \left(\rho_p c_p \frac{2.25 \times 10^3}{[\sigma_{dn_{rup}}]}\right)^{0.5};$$

For a cylindrical charge

$$\sigma_{r_{max}}^{rup} = \rho_p c_p \frac{545}{(\overline{L} - \overline{R}_p)^{1.1}} = \frac{545\rho_p c_p}{(\overline{L} - 50)^{1.1}} = [\sigma_{dn_{rup}}].$$

Hence, $$\overline{L} = 50 + \left(\rho_p c_p \frac{545}{[\sigma_{dn_{rup}}]}\right)^{0.91}$$

If the method of equivalent materials is adopted for modelling, when $\rho g H \gg \sigma_* + p_{atm}$ the second equation of (3.70) is written approximately as

$$\overline{R}_p \approx \left[\frac{4}{3}\frac{p_{00}\overline{R}_0^{-0.75}}{\rho g H \delta_{ed}}\right]^{1/3}, \qquad \dots (3.74)$$

or assuming that geometric similitude is valid and considering the condition for the method of equivalent materials, i.e.,

$$(\overline{R}_p)_N = (\overline{R}_p)_M; \ \delta_{ed_N} = \delta_{ed_M}; \ \sigma_M = \sigma_N/\gamma_c l_c,$$

we obtain from (3.74),

$$\left(\frac{p_{00}\overline{R}_0^{-0.75}}{\rho g H}\right)_N = \left(\frac{p_{00}\overline{R}_0^{-0.75}}{\rho g H}\right)_M$$

$$\left(\frac{p_{00}}{\gamma H}\right)_N (\overline{R}_0^{-0.75})_N = \left(\frac{p_{00}}{\gamma H}\right)_M (\overline{R}_0^{-0.75})_M.$$

From this expression and considering geometric similitude, we obtain the similitude criteria

$$(p_{00})_M = (p_{00})_N \frac{(\gamma H)_M}{(\gamma H)_N} = (p_{00})_N \frac{1}{\gamma_c l_c}, \qquad \dots (3.75)$$

$$(R_{0_z})_N/(R_{0_z})_M = (R_0)_N/(R_0)_M = l_c. \qquad \dots (3.76)$$

Condition (3.75) thus obtained leads to selection of an *EC* having a parameter $(p_{00})_M \ll (p_{00})_N$ in combination with geometric similitude of radii of charge and crater cavity (3.76). This is quite complicated and requires special investigations for developing and producing such *EC*. Because of this, modelling the crushed zone using equivalent materials is currently not in vogue.

3.5 Modelling Patterns of Propagation of Stress Waves and Blast-induced Stress Waves

Let us consider modelling of stress waves within the elastic zone of a blast, which for rocks starts practically at distances ranging from 12 to 15 times the charge radii measured from the centre (axis) of charge and extends to infinity, which is typical for a stress wave in the mid-field zone of a blast and for blast-induced seismic waves in the far-field zone of a blast. The rock massif and its model would be considered as two geometrically similar non-homogeneous elastic media, each of which consists of n homogeneous isotropic fragments of arbitrary shape. The parameters of stress wave (stress σ and displacement u) are taken to be similar if for any pair of similar points at similar time periods the following ratios remain constant,

$$u_N/u_M = u_c, \; \sigma_N/\sigma_M = \sigma_c.$$

Equations of motion of a non-homogeneous elastic medium according to Lamé, conditions of uniqueness (boundary and initial conditions) and constraint equations between stresses and deformations (generalised Hooke's law) are used for establishing similitude criteria and corresponding scales of modelling.

For simplicity, let us consider the propagation of a unidimensional stress wave. In such a case Lamé's equation in vector form for each fragment (layer) would be

$$\rho_i \frac{\partial^2 u_i}{\partial t^2} = (\lambda_i + \mu_i) \text{ grad div } \vec{v}_i + \mu_i \Delta \vec{u}_i, \qquad \ldots (3.77)$$

where i is the index of fragment; $\vec{u}$ the displacement vector; λ_i, μ_i Lamé's constants,

$$\lambda_i = \rho_i (c_{P_i}^2 - c_{S_i}^2); \; \mu_i = \rho_i - c_{S_i}^2, \qquad \ldots (3.78)$$

where c_{P_i}, c_{S_i} are velocity of propagation of longitudinal and transverse waves.

Boundary conditions are written in the form of equality of displacement vectors and conditions of refraction of stresses at all possible boundaries of the non-homogeneous medium under consideration

$$v_q = u_j, \; \sigma_q = \sigma_j k_{q_j}, \; (q \neq j), \qquad \ldots (3.79)$$

where q, j are intermediate indexes of fragments.

Initial conditions are given in the following form:

$$\left.\begin{array}{l} \text{at } t < 0, \; \vec{u}_i = 0, \quad \sigma_i = \sigma_{stat}, \\ \text{at } t = 0, \; \vec{u}_p(r = R_*, t = 0) = \vec{u}(t) \end{array}\right\} \qquad \ldots (3.80)$$

Here the index p indicates the location of source (blast) in the pth fragment of the non-homogeneous medium under consideration.

The relationship between stresses and displacements in the ith layer in vector form would be

$$\sigma_{r_i} = (\lambda_i + 2\mu_i)\frac{\partial u_i}{\partial r}, \qquad \ldots (3.81)$$

$$\sigma_{\varphi_i} = \left[1 - 2\left(\frac{c_{S_i}}{c_{P_i}}\right)^2\right]\sigma_{r_i}. \qquad \ldots (3.82)$$

From equations (3.77), (3.78), (3.81) and (3.82) it is easy to find the similitude criteria mentioned in Sec. 2, for example, by the method of similar transformations. Actually, in place of the quantities contained in these equations and conditions, if the corresponding similitude scales ($l_c, u_c, \lambda_c, \mu_c, E_c$) are introduced and similar transformations made, we get a system of equalities of complexes and simplexes comprising such scales, in particular

from (3.77) $\rho_c u_c/t_c^2 = \lambda_c u_c/l_c^2 = \mu_c u_c/l_c^2$,
from (3.81) $\lambda_c = \mu_c = E_c, \quad \nu_c = 1$,
from (3.78) $\lambda_c = \mu_c = \rho_c v_c^2$,
from (3.82) $\sigma_c = \lambda_c u_c/l_c = \mu_c u_c/l_c$.

Boundary conditions (3.72) indicate that the refraction coefficients k_{qi} and scales of displacements and stresses should be the same for all fragments (layers) of the medium being studied.

From the initial conditions (3.80) before the blast, i.e., at $t < 0$, it is essential for similarity to maintain sameness of stress state in the media up to the moment of blast, i.e., the following conditions must be equal,

$$\left|\sigma_{i_N} = \sigma_{i_M}\right|_{t<0} = \sigma_{stat},$$

$$\left|u_{i_N} = u_{i_M}\right|_{t<0} = 0. \qquad \ldots (3.83)$$

When a blast at a great depth is to be modelled, the first condition from (3.83) is essential, i.e., when there exists a prestressed state of the medium σ_{stat} due to the effect of high rock pressure. To model rock pressure one resorts either to centrifugal modelling or subjects the model to loading under pressure or in a special triaxial compression chamber. Due attention is paid to fulfilment of the spatial and boundary conditions that limit model dimensions, as such conditions often contradict the dimensions and possibilities of the aforesaid loading devices: dimensions of carrier in a centrifuge, press device, hermetic chambers for creating large static pressures etc.

Initial condition (3.80) at $t = 0$ indicates the need for maintaining sameness of blast-induced displacements at a typical distance $R_* > 5$ to $10R_{0_z}$, i.e., fulfilment of conditions

$$\vec{u}_{p_N}(R_{*_N}, t_N) = \vec{u}_0(t_N), \ \vec{u}_{p_M}(R_{*_M}, t_M) = \vec{u}_0(t_M),$$
$$R_{*_N}/R_{*_M} = \vec{u}_0(t_N)/\vec{u}_0(t_M) = l_c, \quad t_N/t_M = t_c.$$

In the breakage of mineral by blasting, this condition should be satisfied at the wall of the charge column or blast hole, whose dimensions (column radius R_K or blast hole R_{bh}) should satisfy the conditions of geometric similitude, i.e.,

$$(R_K)_N = l_c(R_K)_M, (R_{bh})_N = l_c(R_{bh})_M.$$

Thus, for modelling the stress wave in an elastic zone and also for modelling blast-induced seismic waves in a fragment-wise non-homogeneous ideal elastic medium, it suffices if the coefficients of refraction k_{qj}, similitude criteria of Euler, Cauchy and Strouhal number are equal and geometric similitude exists. For this three similitude criteria may be arbitrarily chosen from among the set of similitude scales for quantities governing the propagation pattern of waves under consideration. Usually the following are taken as arbitrary scales: l_c—scale of geometric similitude; ρ_c—scale of density of the medium and c_c—scale of velocities of wave propagation. The remaining governing scales (time t_c and of stress σ_c) are obtained from the similitude indicators of Strouhal number and Euler criterion, and in particular

$$t_c = l_c/c_c, \quad \sigma_c = \rho_c v_c c_c = \rho_c c_c$$

as it is known from many experiments that $\sigma = \rho v c_p$ and $v_c = 1$ in the elastic zone for rocks, owing to the similarity in mechanism of damping of displacement velocity over relative distances in the elastic zone of blast for rocks as well as soils.

The scale of geometric similitude is chosen either from the typical linear dimensions of the model and the full-scale (for example, length of blast holes, LLR, bench height etc.) or from the typical dimensions of explosive charge in the model and the full-scale (for example, equivalent charge radii based on blast energy). The scale of density of the medium is determined by densities of medium of the full-scale and the chosen model. The velocity scale is usually determined by the ratio of propagation velocities of longitudinal waves c_P in the media of the full-scale and the model, i.e., $c_c = c_{P_N}/c_{P_M}$.

From the above features and the derived modelling conditions, it follows that if the medium of the full-scale is used in the model and a charge of the same explosive as used in the full-scale blasted the waves under consideration might very readily be modelled. This would constitute a case of simple geometric similitude. When another type of explosive or other blast source is used, the mass (radius) of explosive charge being modelled is computed based on the concept of energy-based equivalent mass (radius) of charge.

As discussed in [5, 6], numerous complications arise in selecting a medium possessing non-homogeneity and jointing geometrically similar to that of the full-scale.

Other difficulties are also encountered (even in the elastic zone of blast), such as non-linear effects in the propagation of a stress wave in the medium, associated mainly with the non-linear relationship between stress and deformation (deviation from Hooke's law), phenomena of absorption and dispersion of waves—very distinctly exhibited in small-scale blasts. The effect of such factors on modelling of elastic waves was first considered by Vanyagin [5] through an example of a medium with elastic consequence (Boltzman model), for which the dependence of stresses on displacements is expressed as

$$\sigma_r = (\lambda + 2\mu)\left[\frac{\partial u}{\partial r} - \omega \int_0^t h(t-\zeta)\frac{\partial u}{\partial r}\partial\zeta\right], \qquad \ldots (3.84)$$

where ω is a certain constant, characterising damping of amplitude of stress over distance and dispersion of propagation of waves, s^{-1}; $h(t)$ an exponential function of the given medium, given in the form $h(t) = e^{-\alpha t}$; α a certain constant, characterising the exponent of the medium, and also degree of damping of amplitude and duration of compressional phase of stress wave, s^{-1}.

It is readily seen from (3.84) that the following relationships have to be fulfilled in modelling:

$$(R_{0_z}\omega)_N = (R_{0_z}\omega)_M, \quad (R_{0_z}\alpha)_N = (R_{0_z}\alpha)_M,$$

or

$$(\omega/\alpha)_N = (\omega/\alpha)_M,$$

i.e., equivalence of ratios of coefficients ω and α in the full-scale and the model. This would occur when damping of waves attributed to their length in the model and the full-scale is equal, and the function of dispersion of velocities depending on wave frequency represented in a non-dimensional manner, is likewise identical in both. So, under an unaltered geological structure of media for the model and the full-scale, in selecting a scale for geometric similitude one should be guided by the upper limit of wave frequencies at which the structural non-homogeneities and various defects in materials of the medium significantly affect processes of absorption and dispersion of waves. According to the defectoscopy data for monolithic rocks (granite, marble) and materials (concrete, cement, paraffin, mineral wax etc.), the upper limit of frequencies being modelled is correspondingly equal to 1 and 0.5 MHz.

If $f_M = 0.5$ MHz is taken as the highest modelling frequency at which monolithic rocks manifest themselves as a homogeneous medium with an adequate degree of accuracy, then as shown in [5] the maximum scale of geometric similitude l_c, inversely proportional to the frequency scale f_c, in a blast of charge weighing $G_N = 10^6$ kg and recorded vibrations in spalled rocks of 0.03–0.05 or frequency $f_N = 35 - 20$ Hz, would be equal to $G_M = G_N/l_c^3 = 10^6/25^3 \times 10^9 \times 0.064 \times 10^{-6}$ kg $= 0.064$ mg. Hence the weight of the model charge of the same explosive would be $G_M = 0.064$ mg.

Such an evaluation of a model charge is valid only for waves of low amplitudes (sound and seismic waves in the far-field zone of blast in monolithic materials). For a stress wave in the mid-field zone of blast in the same rocks at distances $10 < \bar{r} < 100$, the evaluation given above would be inaccurate because, in that case, the effects associated with dissipative processes and processes of crushing and breakage of rocks should also be taken into account. Therefore the upper limit of model frequencies should be reduced by approximately one order, i.e., $f_M = 5 \times 10^4$ Hz. Then the minimum weight of model charge would still be sufficiently less and equal to $l_c = 2.5 \times 10^3$; $G_M = 0.064$ g.

At such a low charge weight, certain complications arise with measurements of stress wave parameters and conditions of exciting the charge (charge initiation). So, heavier charges, starting from 0.2–0.6 g and more are usually used in modelling. The error in measuring charge weight and also the scatter of blast effect based on shock wave does not exceed 10–15%.

It is to be noted that the presence of cracks in the rock massif may sharply decrease the upper limit of model frequencies and thereby enhance the weight of model charge. Thus at $f_M = 5 \times 10^3$ Hz, the charge weight is equal to 64 g. In this way reduction of model frequency by one order results in enhancement of charge mass in the model by three orders.

To summarise, the stress wave in an elastic zone of blast and blast-induced seismic waves can be readily and conveniently modelled using a full-scale medium. The so-called law of geometric similitude is valid when the similitude scale l_c is unique. Scales of all other quantities are expressed through it and, in particular, time scale $t_c \doteq l_c$, scale of stresses and displacement velocities $\sigma_c = v_c = 1$, scale of forces and power $F_c = N_c = l_c^2$, and scale of work done and energy consumed $A_c = E_c = l_c^3$.

Very small charges, weighing only grams, whose dimensions would be constrained because of reliable and complete detonation, are used for modelling. Modelling may be done either under laboratory conditions or in the field. The latter is generally used for investigating blast-induced seismic waves in the far-field zone of a blast and the former for investigating stress-wave parameters in the mid-field zone of a blast, $10 < \bar{r} < 100$.

To model stress waves in the mid-field zone of a blast under laboratory conditions the method of refracted (in water) wave is the most suitable for studying the positive phase, normal to the measuring plane of the stress wave component in a blast of extended charges and for investigating the positive phase of the radial component in a blast of concentrated charges.

The normal component of stress wave based on the measured pressure in the refracted wave is calculated according to formulae (3.33) and (3.32).

Thus the method of refracted wave in a blast of extended charges enables measurement only of the component of stress wave normal to the measurement plane. Generally, it has a variable value and only in the region of exit at the plane parallel to the charge axis, in the conical part of the wave-front $(D > c_P)$ at distances commensurate with the charge length, is it of the same value and determined by the relationship

$$\sigma_n = \sigma_{r_K} \cos^2 \alpha \sigma_{\varphi K} \sin^2 \alpha,$$

where σ_{r_K} is the stress component directed along the normal to the conical front of the wave; α the angle formed between normal to the conical front and normal to the measurement plane, parallel to the charge axis and determined by the relationship $\sin \alpha = c_P/D$; $\sigma_{\varphi K}$ the stress component oriented along the directrix of conical front of the wave.

The above variability of stress along its front in the process of wave propagation leads to non-linear redistribution of wave energy along its front. As a result, the stress along the front gradually balances and the pyriform front of the wave in the case of $D > c_P$ is transformed into a spherical one. Within the range of relative distances $\bar{r} = 20 - 100$, the redistribution of wave energy and the transformation of its front are not yet completed, however. Therefore, to determine the field of stresses along the entire front of a stress wave, measurements should be taken at several points along the height of the sides of the block, to verify the damping patterns of the wave over distance and patterns of stress distribution around the charge. While doing so, if the wave has a conical front, to evaluate the stress field induced by blasting an extended charge, the refracted wave should be measured at least at 6–7 points along the height of the block: 3–4 points in the region of passage of the conical front (two points should coincide with the exit points at the conjunction of the conical front with the spherical), and the remaining points in the region of exit of the spherical (lower and upper semi-spheres) front.

To obtain a complete wave diagram (phases of compression and tension), as well as the corresponding diagrams of wave components (radial σ_n, tangential σ_φ and vertical σ_z), it is better to take measurements using piezoelectric pressure gauges (according to the methodology discussed in Sec. 3.2.2) on equivalent materials such as acrylic plastic, sand-cement mixture, sodium sulphate etc. To verify the quantitative data of amplitude and duration of wave, normal and radial wave components should be measured by the method of refracted wave in blocks of rock under study at the same relative distances from the charge centre (axis) at which the piezo gauges were installed in blocks made of equivalent materials. Based on such measured data, the coefficients for recalculation of amplitude k_1, of duration k_2

of the stress wave are determined with the objective of extrapolating the experimental results to the full-scale. These coefficients are equal to

$$k_1 = \frac{\sigma_{r,n}^{r.b}}{\sigma_{r,n}^{eq.m}}; \qquad k_2 = \frac{\tau_{+}^{r.b}}{\tau_{+}^{eq.m}},$$

where $\sigma_{r,n}^{r.b}, \tau_{+}^{r.b}$ are maximum amplitude and duration of positive phase r and n components of stress wave respectively, established by the method of refracted wave in a rock block; $\sigma_{r,n}^{eq.m}, \tau_{+}^{eq.m}$ are corresponding quantities determined for r and n components of stress wave in a block made of equivalent materials.

These coefficients, owing to the different physicomechanical properties of equivalent materials and rocks, would depend on the distance from the centre (axis) of the blast. It is necessary, therefore, to plot the experimental relationship of k_1 and k_2 with $\bar{r}$ based on the measured data, and later use it for extrapolating the data obtained in the equivalent material blocks to the rock blocks. Such extrapolation is done according to the relationships:

$$\sigma_{r,n}^{r.b}(\bar{t},\bar{r}) = k_1(\bar{r})\sigma_{r,n}^{eq.m}(\bar{t},\bar{r}),$$

$$\tau_{+}^{r.b}(\bar{r}) = k_2(\bar{r})\tau_{+}^{eq.m}(\bar{r}),$$

$$\sigma_{\varphi}^{r.b}(\bar{r},\bar{t}) = k_1(\bar{r})\sigma_{\varphi}^{eq.m}(\bar{r},\bar{t}),$$

$$\sigma_{z}^{r.b}(\bar{r},\bar{t}) = k_1(\bar{r})\sigma_{z}^{eq.m}(\bar{r},\bar{t}),$$

where $\bar{t}$ and $\bar{r}$ are the time and distance attributed to the equivalent charge radii $R_{0_z}^{*} = (R_{0_z})_M \left[\frac{(\gamma Q)_M}{(\gamma Q)_N}\right]^{\alpha}$ ($\alpha = 1/3$ and $\alpha = 1/2$ respectively in a blast of spherical and cylindrical charges).

Extrapolation of model experiments to the full-scale is done by the law of extended geometric similitude with the scale

$$l_c = \frac{(R_{0_z})_N}{(R_{0_z}^{*})_M} = \frac{(R_{0_z})_N}{(R_{0_z})_M}\left[\frac{(\gamma Q)_N}{(\gamma Q)_M}\right]^{\alpha}.$$

3.6 Modelling Effects of Blast-induced Seismic Waves and Stress Waves on Buildings and Structures

Vibrations acting on different elements of structures may be divided into two groups: (1) natural dynamic effects occurring from earthquakes, rockslides, avalanches etc. Such vibrations are generally termed **Seismic**; (2) vibrations induced by blasts of various nature, including those from blasting operations. These are called **Blast-induced Seismic Vibrations**, as distinct from the first group.

Further, the action of such vibrations is classified as: (1) global seismic effect, when the source of vibrations (earthquake, powerful explosion) is situated far from the study object; (2) localised seismic effect due to blast, when the source of disturbance (local effect of unit or group blasts of small scale) is situated in the immediate vicinity of the study object.

To model the behaviour and damage of such objects, one should know the parameters and nature of action of both stress waves and blast-induced seismic waves (vibrations).

A specific methodology based on the investigations of M.A. Sadovsky is available for computing the intensity of seismic effect of a blast and corresponding safe distances. It is specifically included in Safety Regulations applicable to blasting operations. Localised blast-induced seismic effect is a more complicated phenomenon because associated with the interaction of stress waves generated by firing individual charges or group of instantaneously fired charges, located at different points surrounding the protected object space, at approximately the same distances from it.

3.6.1 Modelling Effects of Stress Waves

Let us consider modelling the effects of stress waves, also termed as local blast-induced seismic effect. This phenomenon is associated with the interference of stress waves due to firing of a series of relatively small charges, situated in the immediate vicinity of the study object (underground workings, pillars, mines, adits etc.). Under the local effect of blast waves, as distinct from the overall seismic effect, the massif may be fractured, allowing the passage of stress waves.

In modelling the local effect of stress waves, one should solve the problems concerning modelling stress waves and behaviour of study objects. These problems are interrelated as the objects situated in the massif themselves affect the stress waves incident on them or passing around them, thereby acting as sources for reflected, diffracted, bent and refracted waves. Such additional waves interact with the incident waves, complicating the overall wave picture, and weaken or intensify the total wave effect. An analytical solution for such an interaction is very difficult and hence modelling is resorted to.

Let us consider determination of modelling conditions through an example of strength evaluation of a simple construction, such as a strip, when a blast-induced stress wave acts on it. In this case its strength is ultimately established by the propagation of elastoplastic waves generated in the strip as a result of refraction and reflection of stress waves incident on it and formation of a diffraction field of stresses around it.

Processes of diffraction, reflection and refraction of stress waves as well as processes of initiation and propagation of elastoplastic waves in construction, depend largely on the construction itself, its configuration and materials, support nodes and other factors. Given the compliant nature of certain constructional elements, additional wave phenomena that complicate significantly the wave picture around the construction occur in rock. In the elements of construction itself, three types of elastoplastic waves form: longitudinal, transverse and surface. Apart from these, multiple reflection and refractions occur which further complicate the overall wave picture.

All such multifaceted and complex phenomena accompanying the effect of a non-contact blast on a construction, may be characterised by the following key governing parameters, for example, effect of an underground blast on a strip of lining over a working of finite dimensions (see Sec. 2.3):

$$P, R_{0_z}, \rho_p, c_p, \sigma_K, \rho_K, c_K, \delta_K, E_p, E_K, \nu, z, z', z'', r, t.$$

Here, the following quantities are additionally introduced: ρ_K, c_K—density and velocity of sound in the lining material; δ_K—thickness of lining strip; z, z', z''—linear deformation of the strip and its derivatives, characterising the angle of rotation and curvature of the strip respectively; σ_K—stress acting on the strip; P—certain resultant load on the strip, proportional to the square of the modulus of geometric similitude (linear scale).

Using the π-theorem, it is possible to obtain the following functional relationship in criterial form for flexure of the strip, by taking R_{0_z}, ρ_p, c_p as the principal quantities,

$$\frac{z}{R_{0_z}} = f\left(\frac{r}{R_{0_z}}, \frac{\delta_K}{R_{0_z}}, \frac{c_p t}{R_{0_z}}, \frac{\rho_p c_p}{\rho_K c_K}, \frac{\sigma_K}{\rho_p c_p^2}, \nu, z', z'', R_{0_z}, \frac{P}{E_K \delta_K^2}, \frac{E_K}{E_p}\right)$$

If geometrical similitude is ensured, the similitude criteria $r/R_{0_z}, z/R_{0_z}$, δ_K/R_{0_z} are assured.

The criterion $\sigma_K/(\rho_p c_p^2)$ characterises the properties of the strip and its surrounding medium; it is automatically fulfilled if the blast occurs in one and the same medium (in model and full-scale).

The criterion $\rho_p c_p/(\rho_K c_K)$ gives rise to the similarity of refraction and reflection of stress waves at the medium strip interface at similar points and is fulfilled automatically if medium and material of constructions of the model and the full-scale are one and the same.

The criterion $\nu =$ const, characterises the elastic properties of the construction material and obviously is automatically ensured if the same materials are used in both constructions.

The criterion $R/(E_K \delta_K^2)$ governs the nature of resultant load, which is usually automatically fulfilled by sameness of external load, geometric similitude, surrounding media and construction materials.

The criterion $c_p t/R_{0_z}$ causes similarity of time-related characteristics and requires that the charge radius be proportional to all the time parameters: periods of natural vibrations of the construction elements, duration of action of forces, time-related nature of the effect (diagram of stress versus time etc.).

These criteria are automatically fulfilled under the condition of geometric similitude, if explosive, construction material and the medium of the full-scale are used in the model, i.e., if the model strip is made from the material of the lining being investigated in the full-scale and is placed in the sample of natural rock. Samples are scaled down according to the scale of geometric similitude by l_c times, and reduced charges of explosives used in the full-scale are taken as charges in the model. If other explosives are used, it is necessary to know in advance the variation in parameters characterising the wave form (duration, time of growth of wave amplitude to the maximum), as these can vary distinctly for different *EC*. Moreover, in such a case one should introduce equivalent radius and weight of charge based on the energy of blast, in accordance with the formulae given in Sec. 2.

The scale of geometric similitude is chosen based on technical possibilities, absence of influence of free faces of rock blocks on the picture of interaction of wave with the study object, and also on the minimum possible weight and dimensions of charge under which it would be completely detonated in a stable manner. In such a case the dimensions of blocks would be large compared to those in the modelling of stress wave parameters, as here a more detailed process of interaction of wave with the object is investigated. The shape of the blocks should be so selected that as far as possible waves reflected from its sides should either not fall at all on the study object, or reach the sides after multiple reflections in a highly attenuated form.

This may be achieved either by means of various diffusers at the block faces or by gluing attachments (plates) to the block faces. Diffusers and attachments are made from the same material as that of the block. The diffuser consists of funnels of different shapes, which reduce the amplitude of reflected wave due to geometric divergence, or due to geometric divergence of multiple reflection. If an attachment is used, its thickness δ is selected such that the intensity of sum-total amplitude of stress at the boundary of block with attachment as a result of the superimposition of reflected and incident waves exceeds the rupture strength of the adhesive (gum) σ_{rup}^{gum}, i.e., from the condition

$$\sigma_{r_m}(W + 2\delta) - \sigma_r\left[(W + 2\delta), \left(\tau_N + 2\delta/c_p\right)\right] \geq \sigma_{rup}^{gum}.$$

This expression, taking into account the relationships for σ_n discussed in Sec. 2, is written in the form

$$1 - e^{-\alpha\frac{\delta}{c_p}} \frac{\sin\left[\beta(\tau_N + \delta/c_p)\right]}{\sin(\beta\tau_N)} \geq \frac{\sigma_{rup}^{gum}(\bar{r} + 2\delta)^{1.1}}{545\rho_p c_p}. \qquad \ldots (3.85)$$

Thickness δ of the attachment may be determined from expression (3.85), assigning the value of σ_{rup}^{gum} by the method of successive approximations or, conversely, knowing the value of δ the required strength of gum σ_{rup}^{gum} can also be obtained.

3.6.2 Modelling Seismic Effects

Let us study the general blast-induced seismic effect and its modelling by *centrifuge*. Obviously, in order to model any vibrations, the model should be subjected to an additional dynamic effect, which could reproduce the blast-induced seismic effect or earthquake.

If the maximum acceleration a of ground vibration, which is a vector quantity, is taken as the principal criterion of the seismic effect then not only the magnitude but also its direction have to be replicated.

The acceleration of vibration is usually expressed as a proportion of acceleration due to gravity. In conformity with the law of geometric similitude, accelerations of vibrations in the full-scale and the model should be related proportionally to $1/l_c$, i.e., $a_N/a_M = v_N t_M/(v_M t_N) = v_c t_c = v_c/l_c = 1/l_c$, since $v_c = 1$. In full-scale conditions the values of a_N vary over a fairly wide range: from thousandth to tenth parts of g_N during earthquakes up to tenth and hundredth parts of g_N for blast-induced seismic waves.

Seismic vibrations are characterised, apart from accelerations, by the time period T, which also varies over a wide range from several hundredth fractions of a second to several seconds. Finally, the key parameters of seismic vibrations are ground displacement u and the rate of displacement v.

Seismic vibrations in the study zone of blast are quasi-harmonic vibrations. Hence the amplitudes of all parameters of interest may be related by the following,

$$v = 2\pi u/T, \ a = 4\pi^2 u/T^2.$$

Arising out of the geometric similitude, we have for these parameters the following scales of modelling:

$$u_M = u_N/l_c,$$

$$T_M = 2\pi \sqrt{u_M/a_M} = 2\pi\sqrt{u_N/(a_N l_c^2)} = T_N/l_c.$$

If the medium is the same in the model and the full-scale, then the propagation velocity c_p is also the same in them and wavelengths $\lambda = c_p^T$ would also be in the same proportion as the displacements, i.e., $\lambda_M = \lambda_N/l_c$.

From the relationships thus obtained, it follows that for modelling the effect of seismic vibrations it is essential that the displacements, periods and wavelengths in the model be less than those in the full-scale to the extent of the modulus of geometric similitude l_c, while the acceleration should be more than that in the full-scale by the same modulus l_c.

Let us consider modelling the effect of blast-induced waves in a centrifuge, i.e., by centrifugal modelling.

Direction of action of acceleration is the principal initial condition to be considered. In the full-scale the damaging effect of acceleration due to seismic vibrations is observed in most cases in a horizontal direction. In centrifugal modelling the horizontal direction (as related to $a_M = g_M$) is usually the most damaging. Let us therefore consider *modelling of horizontal seismic vibrations.*

To reproduce such vibrations the so-called suspended pendulum type probe with model can be used, employing a parallelogram device (Fig. 29,a). The period of vibrations is given by

$$T_M = 2\pi\sqrt{b/g_M},$$

where b is the arm of the parallelogram.

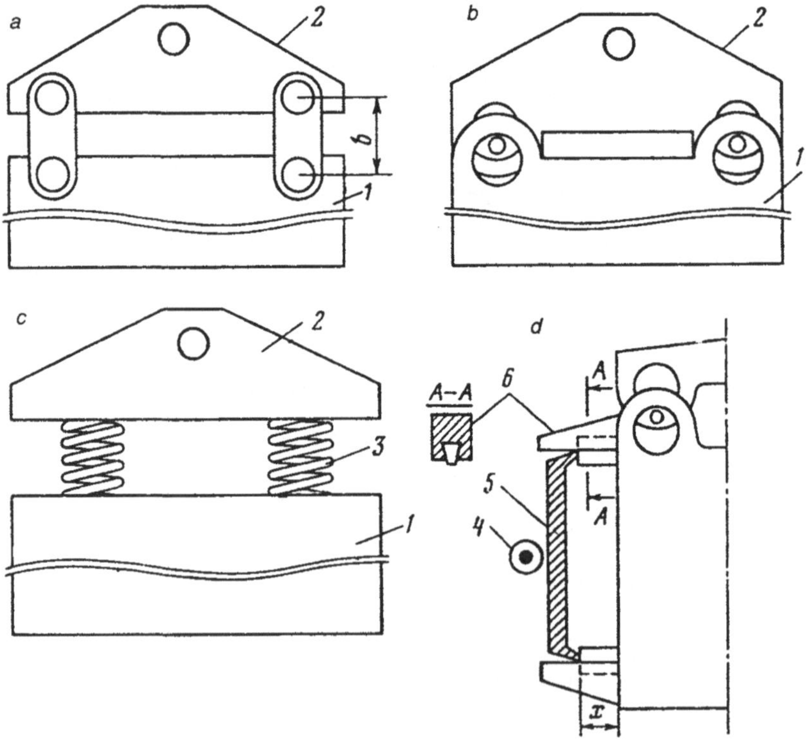

Fig. 29. Methods of suspending buckets in the centrifuge (a, b, c) and bucket vibration by an explosion (d).

1—bucket; 2—suspension; 3—spring; 4—charge; 5—metallic strip; 6—bracket with a gap for inserting strip in the form of a 'forktail'.

Replacing T_M by T_N and g_M by g_N according to the relationship $g_M = g_N l_c$, we get

$$T_M = T_N/l_c = 2\pi\sqrt{b/(g_N l_c)};$$
$$b = T_N^2 g_N/(4\pi^2 l_c). \qquad \dots (3.86)$$

The arm of the parallelogram is selected from condition (3.86). However, if the calculated value of b is less than 2–3 mm, it would be better to use a suspended roller device that provides a complete time period for bucket vibrations (Fig. 29, b). In this case the arm would be quantified by the expression $b = 2(R - r)$, where R is the aperture radius and r the radius of roller inserted into this aperture.

In both cases vertical acceleration a_y should also be considered. Calculations show that its ratio to the horizontal acceleration a_x would be nearer to the ratio $(a_y)_M/(a_x)_M = u_M/b$, but as $u_M \ll b$, then $(a_y)_M$ is negligibly small compared to $(a_x)_M$, i.e., we have $(a_y)_M \ll (a_x)_M$.

If only the vertical component of acceleration is required, spring type suspension of the bucket (Fig. 29, c) is employed. Stiffness of the spring is indicated by the stiffness coefficient K. In this case the period of vibrations of a body weighing M is expressed as

$$T = 2\pi\sqrt{M/K}. \qquad \dots (3.87)$$

In this particular case $M = G_M/g_N$. Then,

$$T_M = T_N/l_c = 2\pi\sqrt{G_M/(g_N K)}, \qquad \dots (3.88)$$

where G_M is the weight of the model inclusive of the bucket.

From equation (3.88), the requirement of stiffness of spring K is specified as follows:

$$K = 4\pi^2 G_M l_c^2/(g_N T_N^2).$$

The bucket with model is suspended from the spring. Due to this suspension, the spring is elongated by a quantity u_{spr} computed from the formula

$$u_{spr} = \frac{G_M(l_c - 1)}{K} = \frac{G_M l_c}{K} = \frac{g_N T_N^2}{4\pi^2 l_c}. \qquad \dots (3.89)$$

Comparing (3.89) with (3.86), it can be seen that these expressions coincide, i.e., $b = u_{spr}$.

To achieve such an oscillatory motion of the bucket, it is usually deflected in advance from its state of equilibrium and later locked. After putting the centrifuge into rotation the stoppers are removed and the bucket begins to vibrate. In the spring type the locking device is so fabricated that when the centrifuge is speeded up, the spring elongates only by the quantity $u_{spr} - u_M$

and not by u_{spr}. Then, when the stopper of the spring is removed, it will vibrate with an amplitude u_M.

In certain cases the locking mechanisms are replaced by devices for active swinging of the bucket while it is revolving. With this in view, a steel strip 10 mm in thickness, on which an explosive charge weighing q is placed (Fig. 29, d), is fastened to the side wall of the bucket (in horizontal vibrations) or at the bottom (in vertical vibrations). On firing, the plate along with the bucket receives a pulse (N.f) equivalent to that of firing a Trotyl charge

$$J \approx 1000\, q \qquad \ldots (3.90)$$

which imparts to the bucket with the modelled energy for movement

$$E = J^2 g_N (2G_M). \qquad \ldots (3.91)$$

To obtain vibrations of specific amplitude, it is necessary that a certain amount of energy be imparted to the bucket as given by

$$E = \frac{Mv_M^2}{2} = 2\pi^2 \frac{u_M G_M}{T_M^2 g_M}. \qquad \ldots (3.92)$$

Equating (3.91) to (3.92) and considering (3.90), we get this expression for the required charge weight

$$g = 2\pi u_M G_M / (T_M g_N 10^3) = \pi u_M G_M l_c / 500 T_N g_N.$$

In this method it may be necessary to soften the initial pulse, as it may be accompanied by unforeseen excessive loading on the model. A shock-absorbing device is introduced between the plate and the bucket to avoid such an effect. This may be made in the form of a mechanism displacing the plate under the action of certain frictional force F by a quantity x, while fulfilling the following conditions

$$F \gg G_M a_M / g_N, \; x \gg 2u_M.$$

When vibrations are initiated in the model by any of the methods described above, they gradually dampen down and this process is usually recorded by seismographs as variation in amplitude of vibrations over time. Preliminary evaluation of damping may be computed taking into consideration the mechanism of pendulum type suspension of the bucket.

Rough computations made by G.I. Pokrovsky made it possible to obtain the following expression for damping of amplitude of vibrations over time

$$u_M / (u_M)_{max} = 1 - \frac{g_N f \frac{r}{b} l_c T_M}{\pi^2 (u_M)_{max}}, \qquad \ldots (3.93)$$

where f is the coefficient of friction in the pendulum type suspension mechanism; r the radius of suspension roller. Here the period of complete damping

t_0 is determined from expression (3.93) by equating it to zero, i.e.,

$$t_0 = \frac{\pi^2 (u_M)_{max}}{g_N l_c f T_M r/b}.$$

Hence the number of vibrations during this period

$$n_0 = t_0/T_M = \pi^2 (u_M)_{max}/(g_N l_c f T_M^2 r/b).$$

The overall seismic effect can thus be modelled by the centrifugal modelling method, if the condition of law of geometric similitude is observed under known parameters of seismic vibrations $(a_N, u_N, v_N, T_N, \lambda_N)$. The modelled study object is placed in the bucket of the centrifuge and subjected to the action of vibrations by the above-described methods, wherein parameters $(a_M, u_M, v_M, T_M, \lambda_M)$ are calculated according to the formula of quasi-harmonic vibrations derived at the beginning of the Sec. 3.6.2.

3.7 Modelling Various Phenomena Associated with the Piston Effect of Gaseous Products of an Explosion

The phenomena associated with the piston effect of blast products comprise the following processes: formation of a crater, zones of compression, crushing and crack formation as well as shear and throw of broken rock, rock fragmentation during throw, its crushing, piling and compaction of ejected rock. The entire process of development of a blast may be divided into stages. In the *first stage* a crater with a radius R_0 is formed, just as in a camouflet blast. In this instance displacement of the free face is very negligible; only slight flaking of individual fragments is discernible. In the *second stage* of breakage, under the pressure of gases the rock in the cavity moves in the direction of LLR, gaining a certain amount of kinetic energy which, in the *third stage,* provides inertial motion to the rock mass. The *fourth stage* involves heaping of fragmented rock.

The last two stages occur under the action of inertial forces because the balance of kinetic energy remains the rock mass in the second stage. These stages mainly determine the piling of rock and the final state of the fragmented rock mass. We shall consider here the last three stages of blast effect on breakage of rock, and in particular: initial movement of broken rock, inertial motion of rock mass, formation of muck pile (blast cone) and compaction of the rock mass.

3.7.1 Modelling Initial Movement of Broken Rock

Let us study the phenomenon of initial movement of broken rock under the pressure of gaseous products of blast through an example of bench blasting of rocks. For simplicity, we shall take a vertical bench, which is typical for hard rocks ($\alpha = 80 - 90°$).

Let us assume that broken rock, due to the action of gaseous products of explosion, begins to move after complete development of the crater as well as formation of a slot between two adjacent charges and that it moves by a short distance along the blasthole line under the action of reflected stress wave (Fig. 30, b). Here the pressure of gaseous products of a blast is calculated by

$$p_t/p_0 = V_m^{\kappa}/(V_m + l_{bh}u_m a)^{\kappa}, \qquad \ldots (3.94)$$

where p_0 is the quasi-static pressure of blast products after the cavity has attained its maximum dimensions R_{0_m}; V_m volume of cavity after its complete development, $V_m = \pi R_{0_m}^2 \times l_{bh}$; u_m maximum displacement of rock along blasthole line under the action of reflected stress wave; a the spacing between blastholes; l_{bh} length of blast hole; p_t quasi-static pressure of explosion products at the initial moment t of rockmass displacement, equivalent to the following value, according to (3.94),

$$p_t = p_0 V_m^{\kappa}/(V_m + u_m l_{bh} a)^{\kappa}.$$

Let us consider the movement of an elementary volume of rock mass m having a cross-section of $S = 1$ m^2 and length equivalent to the LLR, W (see Fig. 30, a):

$$m = \rho_{br} WS = \rho_{br} W,$$

which receives initial kinetic energy for movement E_0, $E_0 = mv_0^2/2$, where v_0 is the initial velocity of the rock mass.

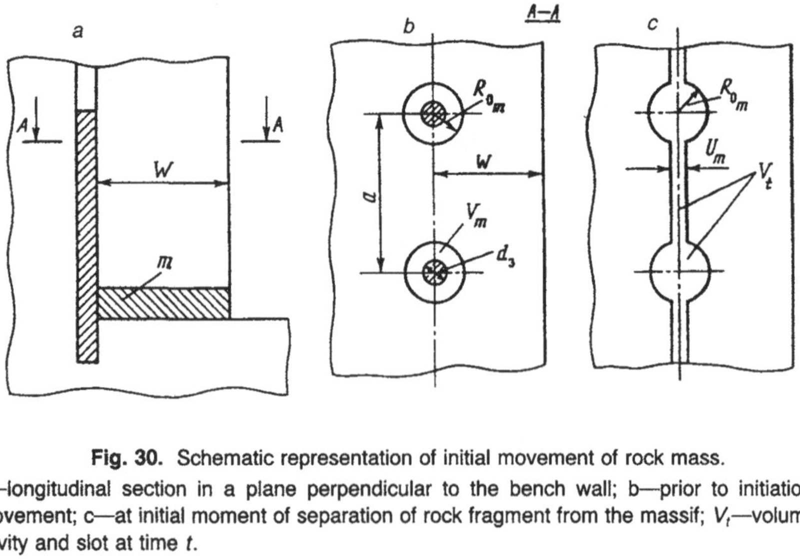

Fig. 30. Schematic representation of initial movement of rock mass.
a—longitudinal section in a plane perpendicular to the bench wall; b—prior to initiation of movement; c—at initial moment of separation of rock fragment from the massif; V_t—volume of cavity and slot at time t.

Let us write the equation of dynamic equilibrium of an elementary mass m, so as to determine the initial velocity of the rock mass,

$$F_1 + F_2 + F_3 + m\frac{d^2u}{dt^2} + \varphi\left(\frac{du^2}{dt}\right)^2 = p_tS = p_t, \qquad \ldots (3.95)$$

where F_1 is the force due to atmospheric pressure, $F_1 = p_{atm}S = p_{atm} = 10^5$ N; F_2 the gravitational force component acting in the direction of LLR, $F_2 = mg\sin\alpha$ (α—angle formed between the normal drawn to LLR and the horizontal plane), equal in the given case to zero, i.e., $F_2 = 0$, as $\alpha = 0$ and $\sin\alpha = 0$; u the current value of displacement of elementary mass m; t current time as reckoned from the moment of firing; F_3 frictional (cohesive) force, $F_3 = \sigma_{compr} \times S = \sigma_{compr}$; φ air resistance constant, $\varphi = \frac{1}{2}C_x\rho_a S$ (where C_x—coefficient of frontal resistance, ρ_{atm}—air density; S—cross-section of elementary mass $S = 1$ m^2); p_tS the force being considered at the moment, exerted by the pressure of explosion products in the crater and in the slot between charges, acting on the cross-section $S = 1$ m^2 of the elementary mass m.

The theoretical and experimental evaluations given in [14] show that air resistance at the beginning of movement is very small; hence the terms in equation (3.95) involving φ may be neglected. In such a case equation (3.95), taking into account $F_2 = 0$, becomes greatly simplified:

$$p_{atm} + \sigma_{coh} + \rho_{br}W\frac{d^2u}{dt^2} - p_t = 0. \qquad \ldots (3.96)$$

The pressure of explosion products acting on an elementary rock mass, according to relationship (3.94) and as per the value of coefficient $\kappa = 1.25$, is given by

$$p_t = p_0(V_{m_0}/(V_{m_0} + u_{atm}))^{1.25}, \qquad \ldots (3.97)$$

where $V_{m_0} = R^2_{0_m}$; $l_{bh} = 1$ m; p_0 the maximum pressure in the cavity at V_{m_0},

$$p_0 = p_{00}\left(\frac{R_{0_z}}{R_{0_m}}\right)^{2\kappa} = 8.5 \times 10^8\left(\frac{R^*_{0_z}}{R_{0_m}}\right)^{2.5}, \qquad \ldots (3.98)$$

where $R^*_{0_z}$ is the charge radius of the explosive being used equivalent to that of Trotyl

$$R^*_{0_z} = R_{0_z}\,(\rho_{EC}Q_{EC}/(\rho_{frac}Q_{frac}))^{0.5}.$$

Expressing acceleration d^2u/dt^2 through velocity v and its derivative dv/du, i.e., $d^2u/dt^2 = v\partial v/\partial u$, and substituting it in equation (3.96) while considering the expression for p_t (3.97), we get

$$\rho_p Wvdv = \frac{p_0(V_{m_0})^{1.25}du}{(V_{m_0} + au)^{1.25}} - (p_{atm} + \sigma_{coh})du.$$

Integrating this equation within limits from v_1 (displacement velocity caused by stress wave) to v_0 (initial velocity caused by the action of explosion products) in the left side and from u_1 to u_m (displacement of rock mass caused respectively by stress wave and quasi-static pressure of explosion products) in the right side, and considering that u_1 is very low (u_1 0), we obtain the equation for determining v_0 as follows:

$$\frac{\rho W}{2}(v_1^2 - v_0^2) = -\frac{p_0}{0.25a}\left[V_{m_0} - \frac{V_{m_0}^{1.25}}{(V_{m_0} + au_m^{0.25})}\right] + (p_{atm} + \sigma_{coh})u_m.$$

Hence, the expression for v_0 would be

$$v_0 = \left\{\frac{8p_0}{\rho W_{atm}}\left[V_{m_0} - \frac{V_{m_0}^{1.25}}{(V_{m_0} + au_m)^{0.25}}\right] - 2(p_{atm} - \sigma_{coh})\frac{u_m}{8W} + v_1^2\right\}^{1/2}. \quad \ldots (3.99)$$

Velocity v_1, imparted due to action of the stress wave, can be roughly evaluated according to the relationship suggested by D.M. Bronnikov, introducing in the expression the value of dynamic tensile strength $[\sigma_{dn_{rup}}]$ in place of static strength, i.e., using the formula

$$v_1 \approx c_p\left[\sigma_{dn_{rup}}\right]/E. \quad \ldots (3.100)$$

Maximum displacement of rock u_m can be determined from equation (3.96), putting the value of acceleration as zero, i.e., $\frac{d^2u}{dt^2} = 0$, and substituting in it the value of p_t as obtained from (3.97).

After transformations, we obtain,

$$u_m = \frac{V_{m_0}}{a}\left[\left(\frac{p_0}{p_{atm} + \sigma_{coh}}\right)^{1/1.25} - 1\right]. \quad \ldots (3.101)$$

Substituting the value of u_m from (3.101) and v_1 from (3.100) in equation (3.99) and taking $m = \rho_p W$, we get an expression for v_0 in the form

$$v_0 = \left\{\frac{8p_0 V_{m_0}}{\rho_p W_{atm}}\left[1 - 1.25\left(\frac{p_{atm} + \sigma_{coh}}{p_0}\right)^{0.2} + 0.25\frac{p_{atm} + \sigma_{coh}}{p_0}\right]\right.$$
$$\left. + c_p^2\left(\frac{\sigma_{dn_{rup}}}{E}\right)^2\right\}^{1/2}.$$

Expressing p_0 through p_{00}, R_{0_m} and R_0^* from (3.98) and V_{m_0} through R_{0_m}, we get the final relationship for determining v_0:

$$v_0 = \left\{\frac{25.13p_{00}(R_{0_z}^*)^2}{\rho_{br} Wa(\overline{R}_{0_m})^{0.5}}\left[1 - 1.25\left(\frac{p_{atm} + \sigma_{coh}}{p_{00}}\right)^{0.2}(\overline{R}_{0_m})^{0.5}\right.\right.$$

$$+0.25\frac{p_{atm}+\sigma_{coh}}{p_{00}}(\overline{R}_{0_m})^{2.5}\Bigg] + c_p^2\left(\frac{[\sigma_{dn_{rup}}]}{E}\right)^2\Bigg\}^{0.5} \quad \ldots (3.102)$$

The third term within square bracket of (3.102) being negligibly small compared to the others, may be ignored and the simplified formula for v_0 written as

$$v_0 = \left\{\frac{25.13p_{00}(R^*_{0_z})^2}{\rho_p Wa(\overline{R}_{0_m})^{0.5}}\left[1-1.25\left(\frac{p_{atm}+\sigma_{coh}}{p_{00}}\right)^{0.2}(\overline{R}_{0_m})^{0.5}\right] + c_p^2\frac{[\sigma_{dn_{rup}}]^2}{E^2}\right\}^{0.5} \quad \ldots (3.103)$$

From (3.103) it follows that, if geometric similitude between the linear dimension scale l_c and time scale $t_c = l_c$ exists, then the displacement velocity scale $v_c = 1$ if the following complexes of quantities both in the model and the full-scale are equal:

$$\frac{(p_{00})_N}{(\rho_p)_N(\overline{R}_{0_m})_N^{0.5}} = \frac{(p_{00})_M}{(\rho_p)_N(\overline{R}_{0_m})_M^{0.5}}, \quad \ldots (3.104)$$

$$\frac{(p_{atm}+\sigma_{coh})_N^{0.2}}{(\rho_p)_N}(p_{00})_N^{0.8} = \frac{(p_{atm}+\sigma_{coh})_M^{0.2}}{(\rho_p)_M}(p_{00})_M^{0.8}, \quad \ldots (3.105)$$

$$\left(c_p\frac{[\sigma_{dn_{rup}}]}{E}\right) = \left(c_p\frac{[\sigma_{dn_{rup}}]}{E}\right)_M. \quad \ldots (3.106)$$

Considering relationship (3.73), condition (3.104) is written as

$$\frac{(p_{00})_N}{(\rho_p)_N(\rho_p c_S^2)_N^{-0.141}} = \frac{(p_{00})_M}{(\rho_p)_M(\rho_p c_S^2)_M^{-0.141}}. \quad \ldots (3.107)$$

From conditions (3.104) to (3.106), it follows that when specimens made from natural rock are used in the model, i.e., at

$$(\rho_p)_N = (\rho_p)_M;\ (c_S)_N = (c_S)_M;\ (\sigma_{coh})_N = (\sigma_{coh})_M;$$
$$(c_p)_N \doteq (c_p)_M;\ [\sigma_{dn_{rup}}]_N = [\sigma_{dn_{rup}}]_M;$$

$E_N = E_M$, the following condition should necessarily be satisfied for modelling

$$(p_{00})_N = (p_{00})_M,$$

which leads to the requirement of selecting an *EC* based on condition (3.46) or (3.47) and solving the cubic equation (3.48) for ρ_{EC}.

Thus to model the initial displacement of a rock mass, the condition of extended geometric similitude may be used by conducting experiments on rock specimens and selecting the *EC* for the model based on formulae (3.47) and (3.48). However, it should be noted that the explosive charge selected for the specified conditions should break the rock specimen into fragments geometrically similar to those in the full-scale blast. This condition can be fulfilled if the specimen has a structure vis-à-vis jointing and non-homogeneity, geometrically similar to that of the full-scale massif being blasted. It is not practically possible to choose such a rock specimen. Generally the specimen would be a monolithic fragment with natural microjointing, dissimilar in orientation and dimensions to that of the macrojointing in the massif. So to break it into smaller fragments approximately similar to the fragments obtained in the full-scale blast, it is necessary to increase the level of stresses, which imposes additional requirements on the model charge, i.e., high shattering effect. These aspects highlight the complexity of modelling the initial movement of rock; in other words, the difficulty in selecting a rock specimen and model charge, which possess on the one hand a high shattering effect and, on the other, such initial pressure of explosion products as to satisfy the conditions in (3.46) and (3.48), resulting in reduced density of model charge as shown by calculations. Such an exercise calls for special investigations.

If the equivalent material method is used, modelling becomes considerably complicated since, in addition to the laborious process of selection of equivalent material, it is essential to select the corresponding *EC* based on conditions (3.104) and (3.105), necessitating again special investigations.

In the modelling of initial movement in throw blasts it is necessary to take into account the gravitational component $F = mg\sin\alpha$ in equations (3.102) and (3.103). For this the condition of modelling (3.105) is written as

$$\frac{p_0 + \sigma_{coh} + \rho Wg)^{0.2_N}}{(\rho_p)_N}(p_{00})_N^{0.8} = \frac{(p_{atm} + \sigma_{coh} + \rho gW)_M^{0.2}}{(\rho_p)_M}(p_{00})_N^{0.8}.$$

If rock specimens are used in modelling, we have:

$$(p_{atm})_N = (p_{atm})_M,\ (\sigma_{coh})_N = (\sigma_{coh})_M,\ (\rho_p)_N = (\rho_p)_M,$$
$$(Wg)_N = (Wg)_M \text{ or } g_M = g_N W_N/W_M = l_c g_N.$$

Hence, it is essential to satisfy additionally the condition of centrifugal modelling $g_M = l_c g_N$, i.e., to artificially increase the weight of the model material by placing the model in a centrifuge or applying load by other methods (for example, sudden arrest of a model falling from a certain height). If the force $F_1 + F_3 > F_2$, then while modelling one can ignore force F_2. In the opposite case, the scale effect comes into play,

$$l_M \approx \left\{\frac{(p_{00})_N^{0.8}[1 - 1.25(\sigma_{coh} + p_{atm} + \rho Wg)_N^{0.2}]}{(p_{00})_M^{0.8}[1 - 1.25(\sigma_{coh} + p_{atm} + \rho Wg)_M^{0.2}]}\right\}^{0.5}, \qquad \ldots (3.108)$$

while the velocity of initial displacement is determined by formula $(v_0)_N = l_M(v_0)_M$ taking l_M from formula (3.108).

Thus the task of modelling the initial movement in the throw of fragmented rock is further complicated by the requirement of similarity of ejected rock mass.

A *combined method of modelling* has been suggested for modelling the initial movement of rock mass v_0 [7]. The main features of this method are as follows. The model massif comprises rock specimens in the form of a core (ball-shaped) with explosive charges placed at their axis (centre). The charge is selected on the basis of equivalence of initial pressure p_{00} in the model and the full-scale, i.e., according to the formulae (3.104) to (3.106), while the equivalent material is chosen in the form of a discrete medium (for example, coarse-grained sand), with median grain size equal to the average fragment of rock mass obtained in the full-scale breakage of rock, taking into account geometric similitude. Dimensions of the rock core (ball) are taken from the condition of geometric similitude of crushing zone; in particular the relative radii of the core (ball) should be equal to the radius of crushing zone in the full-scale blast, while the length of core should be equal to length of model charge plus core diameter.

The relative values of the radii of the crushing zone for concentrated (spherical) $\overline{R}_{cr}^{sph}$ and extended (cylindrical) $\overline{R}_{cr}^{cyl}$ charges are determined by the following relationships [4]

$$\overline{R}_{cr}^{sph} = \left[\frac{3.54\rho_p c_p}{2(1-\nu)[\sigma_{dn_{shear}}]}\right]^{0.455};$$

$$\overline{R}_{cr}^{cyl} = \left[\frac{1.1\rho_p c_p}{2[\sigma_{dn_{shear}}]}(1 + c_1 + c_2 + \overline{R}_{cr}^{cyl})\right]^{0.714}, \qquad \ldots (3.109)$$

and their absolute values by the formulae

$$(R_{cr}^{sph})_M = \overline{R}_{cr}^{sph}(\overline{R}_{0_z}^{sph})_M^*, \ (\overline{R}_{cr}^{cyl})_M = \overline{R}_{cr}^{cyl}(R_{0_z}^{cyl})_M^* \qquad \ldots (3.110)$$

where $\rho_p c_p$ are the acoustic stiffness of rock; $[\sigma_{shear}^{dn}]$ the dynamic shear strength of rock; ν Poisson ratio; $(R_{0_z}^{sph})_M$, $(R_{0_z}^{cyl})_M^*$ equivalent radii of model charges based on the blast energy of TEN ($\rho_T = 15{,}000$ kg/m^3; $Q_T =$ 5860 kJ/kg), determined respectively by formulae (2.23) and (2.24); c_1 and c_2 dimensionless constants dependent on acoustic stiffness of rock and determined by formulae (2.82).

The discrete medium (coarse sand) surrounding the core (ball) with a charge is used as a substitute for broken rock at the initial stage of its displacement. To provide stable bench shape and to obtain proximity between the cohesion coefficient of this medium and corresponding coefficient of fragmented rock in the full-scale, the sand grains are moistened with water or weak clay solution or glycerin. To maintain similitude of mass of the fragmented rock mass ejected by both the model and the full-scale, it is essential that the similitude condition be satisfied, assuming that the rock specimen is used in the model, i.e., from the condition $m_c = l_c^3$. In modelling by the suggested combined method, this condition is written in the form of a relationship

$$\rho_N V_N/l_c^3 = \rho_N V_{ts}/2 + \rho_{dm}(V_M - V_{ts}/2)$$
$$= \rho_{dm} V_M + V_{ts}(\rho_N - \rho_{dm})/2, \qquad \ldots (3.111)$$

where V_N and V_M are volume of rock mass ejected by the full-scale and the model; V_{ts} volume of rock specimen (core, ball); ρ_N and ρ_{dm} density of rock and discrete medium respectively (usually $\rho_N > \rho_{dm}$).

In modelling rock breakage by a group (series) of blasthole charges n, relationship (3.111) is written in the following form, after simple transformations:

$$W_M = \frac{\rho_N W_N}{\rho_{dm} l_c} - \left(\frac{\rho_N}{\rho_{dm}} - 1\right) \frac{\pi}{2} \frac{(R_{cr})_M^2}{H_M a_M} [(l_z)_M + 2(R_{cr})_M], \qquad \ldots (3.112)$$

where W_M and W_N are line of least resistance respectively for the model and the full-scale; H_M, a_M bench height and spacing of charges; $(R_{cr})_M$, $(l_z)_M$ radius of crushing zone and charge length respectively in the model obeying geometric similitude, i.e., $H_M = H_N/l_c$, $a_M = a_N/l_c$, $(R_{cr})_N/l_c = (R_{cr})_M$; $(l_z)_N/l_c = (l_z)_M$.

In the case of a single blasthole charge, formula (3.111) is written as

$$W_M = \left\{\frac{\rho_N}{\rho_{dm}} \frac{W_N^2}{l_c^2} - \frac{\pi}{4}\left(\frac{\rho_N}{\rho_{dm}} - 1\right) \frac{(R_{cr}^2)_M}{H_M} [(l_z)_M - 2(R_{cr})_M]\right\}^{0.5} \qquad \ldots (3.113)$$

For blasting a group of crater charges and a single concentrated (spherical) charge, formula (3.111) is correspondingly represented by the following relationships:

$$W_M = \frac{\rho_N W_N}{\rho_{dm} l_c} - \frac{2}{3}\pi\left(\frac{\rho_N}{\rho_{dm}} - 1\right) \frac{(R_{cr}^3)_M}{H_M a_M}, \qquad \ldots (3.114)$$

$$W_M = \left\{\frac{\rho_N W_N^3}{\rho_{dm} l_c^3} - 0.586\left(\frac{\rho_N}{\rho_{dm}} - 1\right) (R_{cr}^3)_M\right\}^{1/3}. \qquad \ldots (3.115)$$

It is to be noted that the value of W_M determined by formulae (3.112) to (3.115) does not obey the law of geometric similitude but exceeds it. The suggested method of modelling using the combined model allows:

—Greater reliability of dynamic similarity of the near-field zone of blast;

—Considerably weakening of the effect of stress wave on the discrete medium, replacing the already fragmented rock, owing to partial reflection of wave at the interface rock specimen-discrete medium;

—Marked simplification of the preparation and conduction of model experiment compared to other methods, for example, the method wherein a vacuum chamber is used and explosion is substituted by special sources of pressure created by electrical heating of a spherical or cylindrical shell filled with compressed air [1]; or the method in which very complicated preliminary investigations are conducted for developing equivalent material for the model and a new *EC*, the blasting of which should ensure breakage of the model in a geometric granulometric composition similar to that in a full-scale blast [14].

The method of combined modelling suggested here involves the preparation of a model in the following sequential stages:

—The scale of geometric similitude l_c is assigned based on characteristic dimension of the bench (for example, bench height H), taking into account the convenience and possibilities of conducting the experiment under laboratory conditions (usually $l_c = 60$ to 100).

—Knowing the charge radius in the full-scale blast and the scale l_c, equivalent model charge radius based on blast energy is determined by

$$(R^*_{0_z})_M = (R_{0_z})_N / l_c.$$

Using cubic equation (3.48) for the selected *EC* (usually TEN, Hexagene etc.), we determine charge density and, according to the Cooke formula (3.47), the value of detonation velocity D_M and the value of the heat of blast transformation Q_M corresponding to it. Based on these data, using the relationship for equivalent charge radius (3.24), we can determine the true radius of *EC* charge in the model

$$(R_{0_z})_M = \frac{(R_{0_z})_N}{l_c} \sqrt{\frac{(\rho_{EC})_N Q_N}{(\rho_{EC})_M Q_M}}.$$

The dimensions of rock specimen (length and radius of core or radius of sphere) are computed by the relationships in (3.109) and (3.110), substituting in them the data for the full-scale rock

$$\rho_p = \rho_N,\ c_p = c_N,\ [\sigma_{dm_{rup}}] \text{ from } [4]$$

and values of charge radii in the model equivalent to TEN by computing them according to formulae (2.23) and (2.24). By doing so, we get core length $(l_K)_M = (l_z)_M + 2(R_{cr})_M$.

Into specimens prepared with computed dimensions (core—$(l_K)_M$ and sphere—$(R^{sph}_{cr})_M$), holes of radius $(R^{cyl}_{0_z})_M$ and $(R^{sph}_{0_z})_M$ are drilled to lengths of $l^c_h = (l_z)_M + (R^{cyl}_{cr})_M$ and $l^b_h = (R^{sph}_{cr})_M + (R^{sph}_{0_z})_M$ respectively for the core and ball-shaped specimen in which corresponding charges are placed with an already computed density $(\rho_{EC})_M$. The remaining uncharged portion of the holes $l^{cyl}_{rem} = (R^{cyl}_{cr})_M$ in the core and $l^b_{rem} = (R^{sph}_{cr})_M - (R^{sph}_{0_z})_M$ in the ball is either grouted or filled with epoxy adhesive.

Based on the average fragment size of the granulometric composition of rock mass in the full-scale, the grain size in the discrete medium (sand) is chosen. Then the bench model is prepared in a special blast chamber (Fig. 31) having two port lights placed opposite to each other on the side

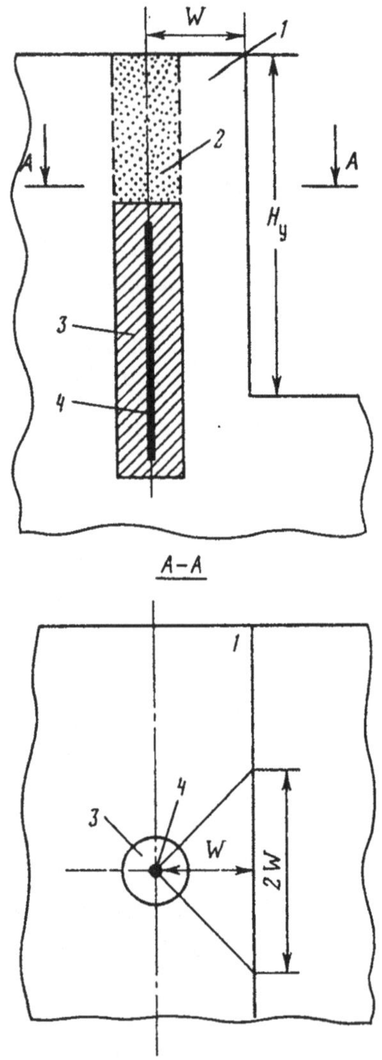

Fig. 31. Modelling the velocity v_0 of initial movement of rock mass by the method of combined models—schematic.
1—discrete medium; 2—stemming; 3—core from a specimen of natural rock; 4—charge; W—LLR; H_y—bench height.

walls of the chamber. The dimensions of the blast chamber are selected with a view to easy placement of the model bench to be blasted in it as well as the number of blast holes. If charges are to be blasted in a row, the chamber width is taken as $b_{ch} \geq a_M(n+1)$, where a_M is spacing between charges, n the number of charges in a row when a single charge is blasted, $b_{ch} \geq 3W_M$.

The length of the blast chamber should not be less than twice its width, i.e., $l_{ch} \geq 2b_{ch}$. If the length of broken rock pile B needs to be measured, the length of the blast chamber would be greater and is determined by the condition $l_{ch} \geq b_{ch} + B_M$, where B_M is the length of pile in the model $B_M = B_N/l_c$. The height of the chamber should not be less than twice the bench height ($h_{ch} \geq 2H_M$). Port lights, made from acrylic plastic or optical glass, are centred in the side walls of the chamber. Their dimensions are 0.5 $l_{ch}\frac{2}{3}H_M$. Illuminating lamps are installed at the top and on one side of the chamber. From the other side of the measuring chamber, a high-speed video camera CKC-1M or very high speed movie, camera SFR, working in the regime of frame-wise photography, are installed. The following frequency of shooting is adopted: 300 frames per second for CKC-1M camera and 250,000 frames per second in the SFR. To indicate the movement of the bench wall over time, a square grid of specific dimensions (5×5 or 10×10 mm) is marked on the glass of the port light oriented towards the movie camera.

3.7.2 Modelling Inertial Motion of Rock Mass

On completion of the *first stage of motion*—displacement of rock mass under the action of stress wave and gaseous products of the blast, the *second stage* begins, i.e., inertial motion of rock mass, consequent to retention of kinetic energy of motion $E_K = mv_0^2/2$ due to initial velocity v_0 and subjection to the action of gravitational forces and air resistance. The motion of the rock mass is described by a system of differential equations, taking into account the opposing actions of the indicated forces, which in rectangular co-ordinates x, y (x-axis oriented horizontally in the direction of motion of the rock mass, y-axis oriented towards a plane in the crest of the bench upwards and perpendicular to the x-axis) is written as

$$m\frac{d^2x}{dt^2} = mg\sin\alpha - \varphi\vec{v}^2\cos\psi,$$
$$m\frac{d^2y}{dt^2} = -mg\cos\alpha - \varphi\vec{v}^2\sin\psi, \qquad \ldots (3.116)$$

where $\vec{v}$ is the velocity vector of elementary mass m:

$$\vec{v}^2 = (dx/dt)^2 + (dy/dt)^2; \;\; \cos\psi = dx/(\vec{v}dt), \sin\psi = dy/(\vec{v}dt);$$

φ the coefficient of air resistance; α the angle of inclination of bench toe to the horizontal plane.

The following initial conditions are imposed on these equations

$$x|_{t=0} = 0, \qquad y|_{t=0} = h;$$

$$\frac{dx}{dt}|_{t=0} = v_0, \qquad \frac{dy}{dt}|_{t=0} = 0,$$

where v_0 is the initial velocity of elementary mass m; h the height at which mass m is located relative to the toe of the bench.

Putting the values of $m = \rho_p WS$ and $\varphi = c_s\rho_{air}S/2$ in equation (3.116) and after transformation we get the following

$$\frac{d^2x}{dt^2} = g\sin\alpha - \frac{-c_s\rho_{air}}{2\rho_p W}\vec{v}\frac{dx}{dt},$$

$$\frac{d^2y}{dt^2} = -g\cos\alpha - \frac{-c_s}{2}\frac{\rho_{air}}{\rho_p W}\vec{v}\frac{dy}{dt}.$$

where ρ_p and ρ_{air} are density of rock and air respectively; S a cross-section of the elementary mass m.

From these equations, by the method of similar transformations, we obtain the following equalities constituted from scales of quantities

$$l_c/t_c^2 = g_c = (\rho_{air})_c v_c l_c/((\rho_p)_c l_c t_c) = (\rho_{air})_c v_c^2/((\rho_p)_c l_c).$$

Equating the second and last terms of these equalities, we get,

$$g_c = (\rho_{air})_c v_c^2/((\rho_p)_c l_c) \text{ or } (\rho_{air})_c v_c^2/((\rho_p)_c g_c l_c) = 1. \qquad \ldots (3.117)$$

If the extended law of geometric similitude is used in modelling, then $l_c = t_c$, $v_c = 1$; if rock specimens are used as the model, i.e. $(\rho_p)_c = 1$ and $(\rho_{air})_c = 1$, then from the similitude indicator (3.117) we get $g_c l_c = 1$ or, expanding it for the full-scale and the model

$$g_N l_N = g_M l_M \text{ and } g_M = g_N l_N/l_M = l_c g_N. \qquad \ldots (3.118)$$

This relationship is the primary condition for centrifugal modelling.

If we reject centrifugal modelling and take $g_c = 1$, and if geometric similitude exists with the arbitrarily chosen scale l_c and the model is prepared from a rock specimen $(\rho_p)_c = 1$, then from (3.117) we obtain the indicator:

$$(\rho_{air})_c v_c^2/l_c = 1 \text{ or } (\rho_{air})_N/(\rho_{air})_M (v_N/v_M)^2 = l_c,$$

revealing that a relationship exists between the scale of velocity v_c and velocity in the model v_M:

$$v_c = \sqrt{l_c/(\rho_{air})_c}, \quad v_M = v_N\sqrt{(\rho_{air})_N/((\rho_{air})_M l_c)}.$$

It follows that for $v_c = 1$ and $v_N = v_M$ it is necessary that $(\rho_{air})_c = l_c$ or $(\rho_{air})_M = (\rho_{air})_N/l_c$, i.e., air density in the model should be lower by l_c times than air density in the full-scale, which means that modelling should be done in a vacuum chamber. The same holds good even in the methodology for a combined model, as in this method due to the increase in LLR, m_c is taken as equal to l_c^3, which occurs for $\rho_c = 1 (m_c = \rho_c l_c^3 = l_c^3)$.

Moreover, in all the cases considered above it is necessary to take into account the condition for modelling the initial velocity of (rock) movement, i.e., the condition $(v_0)_N = (v_0)_M$, which is valid if conditions (3.46) and (3.47) are satisfied, which requires that an *EC* with a density calculated as per equation (3.48), be selected.

It should be noted that in modelling on a centrifuge (centrifugal modelling), Coriolis (complementary) acceleration would affect the movement of fragments, thereby distorting movement. Due to bending in the trajectory of motion, a transverse displacement of points in the trajectory of motion (deviation) occurs in the plane of rotation. This displacement is approximately

$$\delta = [L^2/(2v)]\sqrt{l_c g_N / R_{cen}} = [L^2/(2v)]\sqrt{g_M / R_{cen}}, \qquad \ldots (3.119)$$

where L is the extent of throw of fragments; v the velocity of a body moving due to inertia; R_{cen}—radius of centrifuge.

Accordingly, the level of accuracy of replicating in the model the movement of fragments as seen in the full-scale, considering expression (3.119), is given by

$$\delta/L = [L/(2v)]\sqrt{l_c g_N / R_{cen}} = [L/(2v)]\sqrt{g_M / R_{cen}}.$$

This relationship enables one to assess the accuracy of modelling and how it is distorted by Coriolis acceleration. This equality further enables one to determine the conditions under which a specific accuracy in modelling can be obtained, i.e., the specific value $\delta/L \ll 1$. Based on this relationship, the following parameters can be established.

Upper limit of scale of modelling

$$(l_c)_{max} = 4(\delta/L)^2 v^2 R_{cen}/(L^2 g_N).$$

Upper limit of trajectory length in the model

$$L_{max} = 2(\delta/L)v\sqrt{R_{cen}/(l_c g_N)}.$$

Lower limit of radius of centrifuge

$$(R_{cen})_{min} = L l_c g_N/[4v^2(\delta/L)^2].$$

These conditions need not be compulsorily fulfilled. Based on the experimental results, appropriate corrections can always be made. Thus in the motion picture recording of the trajectory of motion, it may be transformed by applying a correction that excludes the effect of Coriolis acceleration. It should be borne in mind here that the trajectory due to correction should shift in the same direction in which the bucket is rotating. The curvature of trajectory of particle motion in the front part of the bucket (relative to the direction of rotation) should reduce while that in the rear portion increases.

In order to reduce effect of Coriolis acceleration on the trajectory of motion, one more requirement may be imposed, which in an elementary form reduces to the following: during the flight of rock fragments the angle of rotation of centrifuge should be much less than 1 radian. Then we can take the angle of rotation of centrifuge β expressed in radians during the free flight of rock fragments as the measure of error in the result obtained in the modelling of a blast intended for rock casting

$$\beta = \sqrt{L_M / R_K},$$

where L_M is the extent of throw of fragments in the model, R_K the distance of the corresponding point in the bucket from the rotation axis of the centrifuge.

Based on the requirement imposed to the angle of rotation, we have the condition

$$\sqrt{L_M / R_K} \ll 1 \text{ or } L_M / R_K \ll 1; \; L_M / R_K = \beta^2.$$

This is the additional requirement to dimensions of the model being placed in the centrifuge, which can be fulfilled in practice. Hence modelling a blast for rock casting using a centrifuge is feasible in principle. If the accuracy obtained in modelling is insufficient, then appropriate corrections may be introduced while conducting the experiments, i.e., the picture of casting (fly rock) should be deformed by rotating the obtained directions of rock throw to an angle β in the direction of rotation of the centrifuge.

As can be seen from the above, modelling the inertial motion of a broken rock mass is a more difficult process than modelling the initial velocity of movement. This arises from the necessity of considering the influence of gravitational force and resistance of the medium surrounding the rock mass (air, water—when a blast occurs in an unconfined space and in a mined-out area filled with water respectively).

Should centrifugal modelling be adopted using a small rock specimen as the model, rock breakage by a blast into fragments similar to those in the full-scale, as already shown in Sec. 3.7.1, cannot be realised in practice. Hence this method gives a large error of observation due both to indirect measurements and the presence of Coriolis acceleration, which distorts the picture of the muck pile or rock casting. The adoption of a combined model (rock specimen in the form of core or ball, surrounded by a discrete medium sand) needs, according to conditions (3.118) and $v_c = 1$, a reduction in density of the medium surrounding the model (air or water) by the magnitude of geometric scale, while preserving the particular condition for rock mass $m_c \approx l_c^3$. In this case the initial velocities of motion in the model and the full-scale would be equal, but the model experiment should be conducted in a vacuum chamber having $(\rho_{air})_M = (\rho_{air})_N / l_c$ for modelling breakage on a free face. Rock breakage in a mined-out area filled with water should be modelled in a compressed air chamber, in which the air pressure should be

$(\rho_{air})_M = (\rho_{dw})_N / l_c$, where ρ_{dw} is density of water. Here the breaking effect of displaced water falling from above is not taken into account.

3.7.3 Modelling Throw of Rock Mass and Formation of Blast Cone

To study the formation process of a blast cone, we take the following as parameters: H—depth of emplacement, R_0—radius of cavity, p_0—pressure exerted by gases; k—adiabatic index and E_{PB}—gaseous energy. Rock mass is characterised by density ρ_p, cohesion σ_{coh} and coefficient of internal friction f. The governing parameters must also include g—acceleration due to gravity, characterising the gravitational force of ejected rock mass. The unknown parameters of the cone would be: radius R_{un} and height H_{un}. Then the functional relationship for radius of the cone, for example, is written as

$$R_{un} = \Phi(H, R_0, p_0, E_{PB}, \rho_{br}, \sigma_{coh}, k, f, g)$$

which, based on the π-theorem, may be represented in a dimensionless form

$$R_{un}/H = \Phi(R_0/H, E_{PB}/(\rho g H^4), \\ E_{PB}/(p_0 H^3), E_{PB}/(\sigma_{coh} H^3), k, f). \qquad \ldots (3.120)$$

It follows from this expression that geometric similitude may be valid if weight, i.e., the criterion $E_{PB}/(\rho g H^4)$ is neglected. This can be done if the quantity $\rho_p g H$ is small compared to p_0 or σ_{coh}, which is usually the case when light-weight charges are blasted. As a result of processing the experimental data of relatively small-scale blasts, a series of approximate formulae (in particular the formula of Boreskov) was obtained for determining the charge weight required for the formation of a cavity of specified radius. These formulae are written in the form of relationship

$$G/H^3 = C_f(R_{un}/H),$$

where C is a constant dependent on the properties of soils (rock).

In the inverse relationship of quantities $\rho_p g H$ and p_0, the unique governing criterion containing energy term $E_{PB}/(\rho_p g H^4)$ remains. Consequently, in this case the (blasting) cones (equivalence of ratios R_{un}/H) happen to be similar if

$$(E_{PB})_N/(\rho_p g H^4)_N = (E_{PB})_M/(\rho_p g H^4)_M,$$

i.e., when energy of gas in the cavity would increase proportionally to H^4 and not to H^3 (as in the case of geometric similitude). It follows from this that, in a fixed medium, the role of the criterion $E_{PB}/\rho_p g H^4$ would enhance with increase in the scale of the blast. This is confirmed by experimental data obtained from large-scale throw blasts.

G.I. Pokrovsky developed a relationship for large-scale blasts:

$$G_{un} = \frac{\gamma}{2800} H^{7/2}(1 + n_0^2)^2, \qquad \ldots (3.121)$$

where γ is rock density; n_0 the index of rock casting, $n_0 = R_{un}/H$.

This formula has been confirmed by the results of both large-scale and small-scale blasts. It follows from the formula that for modelling it is essential to satisfy the condition $G_1/G_2 = \gamma_1/\gamma_2(H_1/H_2)^{7/2}$ and, further, to comply with geometric similitude.

Another relationship, closer to equality (3.120), also exists:

$$G = \rho_p B \sqrt{\frac{2}{3}} g H^{7/2}(1 + n_0^2)^2, \qquad \ldots (3.122)$$

where g is acceleration due to gravity; ρ_p rock density; B constant dependent on the specific energy of EC; R_{un} the unknown radius of (blast) cone.

Formula (3.122) for the full-scale and the model, when geometric similitude exists, is written as follows:
for the full-scale

$$G_N/H_N^3 = (\rho_p)_N B \sqrt{\frac{2}{3} g_N H_N}(1 + n_0^2)^2,$$

for the model

$$G_M/H_M^3 = (\rho_p)_M B \sqrt{\frac{2}{3} g_M H_M}(1 + n_0^2)^2. \qquad \ldots (3.123)$$

Evidently, in this case when we make a transition from the full-scale to the model, the quantities n_0, B and G/H^3 do not vary. Consequently it follows from condition (3.123) that

$$g_N H_N = g_M H_M, \; g_M = g_N H_N/H_M = g_N l_c,$$

in other words, the principal condition for centrifugal modelling.

However, another method of modelling is also possible. For example, in the model experiment the values of p_0, σ_{coh} and E_{PB} can be reduced so that the dimensionless criteria in (3.120) coincide with those in the full-scale. A vacuum chamber half-filled with dry sand was used in [1] to achieve this reduction. In this chamber at a depth H_M from the surface, a thin spherical balloon filled with air was placed. When heated, the balloon burst and the released bubbles of air ejected sand with a pressure $(p_0)_M$, thereby creating a blast cone. Pressure in the vacuum chamber and in the gas bubble were chosen based on the second and third criteria (3.120); sand strength was chosen in accordance with the fourth criterion and radius and depth of emplacement of sand in accordance with geometric similitude or from equality of the first criterion (3.120). Here, as a first approximation, the adiabatic index K and coefficient of friction f were assumed to be constant, which obviously introduces an error in the results of modelling. This method of modelling was adopted for studying the dependence of crater formation and its dimensions on the depth of charge emplacement for alluvial and loamy soils.

For modelling crater formation the method of combined modelling may also be adopted by placing the model in a vacuum chamber with $(\rho_{an})_M = (\rho_{an})_N / l_c$ and also applying load on the charge by means of a ball-type specimen at a depth $H = W$(W is computed by (3.115)), wherein the scale of geometric similitude is assigned arbitrarily, commensurate with the dimensions of the vacuum chamber as well as the crater, calculated by using either formula (3.121) or (3.122), i.e., from the condition,

$$(R_{un})_N / l_c \leqslant 4L_K,$$

where $(R_{un})_N$ is the radius of the crater in the full-scale, calculated according to formula (3.121) or (3.122); L_K the minimum size of vacuum chamber (length, height, radius etc.).

In this case the scale of geometric similitude is given by

$$l_c = (R_{un})_N / (4L_K).$$

After selecting the scale in such a manner, the type and characteristic of model charge $(\rho_{EC})_M, (R_{0_z})_M$, radius of ball-type rock specimen in which the charge is placed [see formulae (3.109) and (3.110)] and the depth of charge emplacement [see formula (3.115)] are determined according to the methodology discussed in Sec. 3.7.1.

3.8 Modelling the Muck Pile and its Consolidation

Let us examine modelling of a rock pile and its consolidation during a blast, in accordance with the investigations conducted by G.M. Pokrovsky. Obviously, when blasted rock is ejected, it gets broken into individual fragments, which in turn attain a certain velocity. These fragments, while preserving a particular fraction of the velocity, denoted as v_K, fly up to the ground surface (or up to some obstruction) and come to rest. While doing so their kinetic energy $mv_K^2/2$ is spent on further fragmentation (crushing) at points of mutual contact. The finer fragments thus formed fill the pores between coarser fragments. This process continues until the entire reserve of kinetic energy is expended.

The process described above results in the formation of a muck pile which, under certain conditions, would be specified by a mechanical composition that corresponds to a particular minimum value of porosity.

Ultimate squeezing strength σ_* for rock piles would be characterised by the porosity ε of this pile. Thus, at the maximum porosity ε_{max}, the value of $\sigma_* = 0$ and conversely at $\varepsilon = 0$, σ_* increases up to the value of ultimate strength $(\sigma_*)_{max}$ of the rock in the massif and later, as a result of being crushed, the muck pile under consideration is obtained.

In the first approximation, let us assume that σ_* varies uniformly as per linear law from zero to $(\sigma_*)_{max}$ if ε decreases from ε_{max} to zero. In such a case the consolidation energy of the muck pile, when porosity varies from ε_{max} to ε, for a unit initial volume of rock pile is given by

$$E_1 = -\int_{\varepsilon_{max}}^{\varepsilon} \sigma_{max} \left[d\varepsilon - \frac{\varepsilon d\varepsilon}{\varepsilon_{max}} \right] = \frac{\sigma_{max}}{2\varepsilon_{max}} (\varepsilon_{max} - \varepsilon)^2. \qquad \ldots (3.124)$$

The kinetic energy of a unit rock (in its original state before the onset of ε_{max})

$$E_1 = \rho(1 - \varepsilon_{max}) v_K^2/2; \qquad \ldots (3.125)$$

where v_K the collision velocity of fragments during formation of the rock pile. Equating (3.124) and (3.125), we have

$$\rho \left(\frac{1 - \varepsilon_{max}}{2} \right) v_K^2 = \frac{(\sigma_K)_{max}}{2\varepsilon_{max}} (\varepsilon_{max} - \varepsilon)^2$$

or

$$\varepsilon = \varepsilon_{max} - \sqrt{\frac{\rho(\varepsilon_{max} - \varepsilon_{max}^2)}{(\sigma_K)_{max}}}\, v_K.$$

If it is necessary in a model to obtain the value of ε equal to that in the full-scale, this can be realised under the condition of similarity of granulometric composition of rock mass in the model and the full-scale and equivalence of corresponding velocities of rock fragments. The latter is possible if the vertical components of velocity, as determined by gravitational force, coincide. This can be achieved by artificial weightage of fragments in the model, by adopting the method of centrifugal modelling, since acceleration due to gravity g is contained in the expression for this velocity, i.e.,

$$v_z = \sqrt{2gh}.$$

A particular case of pile formation is its creation when rock fragments fall from a certain height H. The velocity of fall $v = \sqrt{2gH}$. It follows from this expression that for equality of velocities in the full-scale and the model, it is essential to comply with the equality $g_N H_N = g_M H_M$ (or $g_M = g_N \frac{H_N}{H_M} = g_N l_c$) which is the principal condition for centrifugal modelling.

It should be noted that exact modelling of rock mass piling is not possible, as this largely depends on the granulometric composition of the rock mass, which is determined primarily by the jointing and non-homogeneity of the rock mass being blasted. In actuality, selection of a model (rock specimen) with similar jointing or type of heterogeneity is not feasible. That is why an artificial model massif consisting of glued (bonded) individual fragments, roughly similar to average fragments obtained in full-scale blasts, is sometimes prepared. For fragmenting and moving such a model, it is also essential to determine the source of blast adequate for obtaining (rock) movement in the full-scale. This is associated with modelling breakage of the model, initial motion of the rock mass and its displacement, i.e., modelling the processes considered earlier in Sec. 3.7. Such an approach to modelling a muck pile requires special investigations on the preparation of model and choice of blast source. Furthermore, to obtain uniform porosity in the muck pile in the full-scale as well as in the model, it is essential to maintain an identical relationship of porosity of muck pile versus strength of rock fragments. This is possible by either weighting the fragments of full-scale rock by adopting for example, the centrifugal modelling method, or by choosing fragments made of equivalent materials possessing lower strength than the strength of full-scale rock.

3.9 Modelling Fragmentation of Rock by Blasting

Rock fragmentation by blasting is a very complicated process and is determined by the action of stress waves as well as the effect of gaseous products of detonation. The degree of fragmentation is considerably affected by the rock structure, mainly by its non-homogeneity and jointing. Quality of fragmentation depends on many factors but especially on these: blasting method, type of explosive, charge diameter, charge configuration, delay intervals, strength of rock, its structural features, non-homogeneity and jointing. As of today, these factors cannot be considered analytically. Hence researchers resort to modelling the process of rock fragmentation. But modelling is a complicated task, dependent on such difficult-to-model parameters as jointing and involving, moreover, a set of sequential and separate procedures.

The process of rock fragmentation may be outlined as follows. First a blast-induced stress wave acts on rock. This wave, during its propagation to and from the free face (as a rarefied wave), creates new cracks and widens already existing ones in both radial and tangential directions. As a result, a network of cracks develops in the rock located between the charge and the free face and under the action of quasi-static pressure of gases the rock disintegrates, forming a rock mass of specific size which, under the effect of products of explosion, moves towards the free face. During this movement the rock fragments collide with each other and break into smaller pieces. Furthermore, during the inertial rock mass motion, mutual collision of fragments and their pulverisation may also occur. The final stage in fragmentation occurs when the rock fragments fall on the bench floor in an open-cast mine or on the floor of workings in an underground mine, whereby a muck pile is formed.

It is obvious from this outline that the first two stages of the process caused by the action of stress wave and piston effect of gaseous products of explosion, exert a decisive influence on the quality of fragmentation. The initial size of a rock mass depends on the natural micro- and macro-jointing of rock, whose modelling is not feasible.

So, two approaches to modelling the process of rock fragmentation are currently available: the first is based on the similitude of elastic wave motions, the second on adoption of the method of equivalent materials. Both approaches are approximate and formulated under certain assumptions, namely

1. That the medium remains elastic right up to the moment of breakage during the process of loading. Obviously, fragmentation of brittle spalling rocks like quartz, quartz glass and other rocks containing a large percentage of quartz, corresponds closely to this assumption. The assumption is not valid for semi-spalling rocks and soils.

2. That the processes of breakage and fragmentation occur instantaneously and their intensity can only be determined by the magnitudes of maximum stresses. This assumption encounters some objections as it does not take into account such important factors as variation and redistribution of stresses during the breakage process, nor influence of pulse and shape of stress wave which vary considerably when the geometric scale of modelling is altered.

It is possible to model the process of fragmentation by introducing these assumptions and under the conditions of geometric and time similitude, similitude of stress fields and strength characteristics of rocks. When the medium used in the model is from the full-scale, modelling of the fragmentation process boils down to modelling of stress waves, i.e., to the extended law of geometric similitude, when the time and linear scales are equal to each other and the velocities at corresponding spatial points and at corresponding time periods are likewise equal.

The main demerits of this approach are: effect of shape and pulse of stress wave on the fragmentation process is not taken into account; it is not possible to comply with the geometric similarity of micro- and macrocracks, which largely determine the granulometric composition of the rock mass in the model and the full-scale.

The *second approach* is associated with the method of equivalent materials, discussed in Sec. 3.1.2. The method has typical complications and demerits. The major disadvantage of the first approach also applies here, namely, the impossibility of modelling non-homogeneity and jointing of the full-scale medium, which are the governing factors for the similarity of granulometric composition of the rock mass. Consequently, this approach is a very approximate and imperfect one. Thus, we can only talk about approximate and partial modelling of rock fragmentation by blasting. For example, in model blasts on blocks of any rock it is possible to study the influence of diameter or charge configuration on the breakage of the block. The comparative results obtained enable one to judge somewhat the influence of these parameters in a full-scale blast. However, while extrapolating model experiments to the full-scale, one should bear in mind the structural condition of the rock, its non-homogeneity and jointing. As this cannot be realised in models, we cannot obtain quantitative results on rock fragmentation based on model experiments.

The first approach is used for approximate modelling of the rock breakage process with the help of an analogue machine EGDA, i.e., it is used in mathematical modelling of rock breakage. It is elaborated in Chapter 4.

The second approach, using equivalent materials, was first suggested in 1961 by Prof. V.R. Imenitov and used by him for modelling the fall (cave-in) and discharge of ore. He assumed that similar fragments should be located in a particular manner in the caved rock, if similarity of acting forces was complied with and if broken fragments in the model were geometrically similar to the fragments in the full-scale. Since during distribution of rock fragments in a muck pile gravitational forces play a key role, whose similarity conditions are too inadequate to ensure similar breakage, it is suggested that a model be created from geometrically similar fragments of a strong material bonded with a low-strength cement. When such a model is blasted, it breaks at the points of binding of fragments, which being freed become distributed in the muck pile under gravitational forces. The scale of charge mass q_c can be computed from the ratio

$$q_c = l_c / K_c \qquad \ldots (3.126)$$

and the scale of charge diameter $(d_z)_c$ from the relationship

$$(d_z)_c = l_c^{3/4} / \sqrt{K_c(\gamma_{EC})_c}, \qquad \ldots (3.127)$$

where l_c is the geometrical scale; $(\gamma_{EC})_c$ scale of explosive density; K_c scale of coefficient of influence of efficiency, charge configuration, strength and density of the medium.

It should be noted that in the given case some uncertainty exists in the selection of scale K_c, which includes the properties of charge and rock. This method can only be used for the process of collapse and distribution of fragments in the muck pile, i.e., for solving a special problem.

There is presently a trend towards adopting the method of equivalent materials for modelling even other specialised problems associated with exploring the effect of blasting conditions on the quality of fragmentation. Based on data on the shape of muck pile and granulometric composition of fragmented rock in a blast in full-scale conditions under specific conditions of blasting (blasthole grid, continuous charge configuration and row-wise firing scheme), the size of material in the model is selected considering the geometric scale l_c. In most cases the model is assembled from equal fragments geometrically similar to the average size of fragments obtained in the full-scale blast. The scales of weight and diameter of charge are chosen according to relationships (3.126) and (3.127). A set of preliminary model blasts are conducted under blasting conditions similar to those in the full-scale with weight and diameter of charge varied until a similarly shaped muck pile and granulometric composition are obtained. Based on this data, scales K_c and $(\gamma_{EC})_c$, constituting part of relationships (3.126) and (3.127), are corrected. Later, the effect of one of the parameters of blasting conditions (for example, blasthole diameter, blasthole grid etc.) on fragmentation is studied by varying this parameter within specific limits to obtain an optimal degree of fragmentation. The optimal conditions of blasting thus obtained are subsequently transferred to the full-scale by taking into account the geometric scale, scales of weight and diameter of charges according to relationships (3.126) and (3.127) with the corrected values of scales K_c and $(\gamma_{EC})_c$.

It should be noted that this method of modelling can only qualitatively evaluate the various methods used for controlling the blast effect on rock breakage. However, here one cannot establish the quantitative variation in the intensity of fragmentation in the full-scale depending on the adopted method of controlling blast effect based only on the data of model blasts. In other words, the granulometric composition of rock mass obtained in a model blast cannot be transferred to the granulometric composition in a full-scale blast, even taking into account the geometric scale. We can only state the extent to which (in percentage) the granulometric composition would vary depending on variation in particular parameters in the conditions of blasting.

Thus, modelling the quality of rock fragmentation in its entirety is very difficult and cannot be solved practically. Only an approximate and partial modelling of this phenomenon is possible by the above indicated methods. The possibilities of the method of equivalent materials are not completely exhausted. During its further development it may be widely used for investigating individual indexes of rock breakage by blasting.

It should also be highlighted that the final granulometric composition of a rock mass is determined after the formation of a muck pile. Therefore, all the conditions and requirements for modelling a muck pile, discussed in Sec. 3.8, should necessarily be considered for modelling rock breakage.

4

Mathematical Fundamentals of Modelling Blast Effects in Rocks

In addition to physical modelling, *mathematical modelling* has been adopted more and more in recent years. In physical modelling a process of direct interest to the researcher is studied on a reduced (or enlarged) model made from either the medium of full-scale or equivalent materials, while in mathematical modelling a system of differential equations describing such a process is studied and solved with the help of special analogue computers (ABM). Any phenomenon (process) that can be described mathematically and can be related to an experiment with a specific physical system structured in such a way that its mathematical description matches with that of the phenomenon (process) under study, can be modelled mathematically. Here the solution using ABM is termed modelling while the analogue computer itself is the model. The possibility of creating such ABM models is determined by similarity in mathematical description of a set of phenomena differing physically, i.e., by the mathematical isomorphism of various physical systems. Generally speaking, any physical system can be termed an ABM wherein specific mathematical relations exist between the continuously varying physical quantities similar to those for the physical system under study or the mathematical problem being solved.

4.1 Theoretical Principles of Mathematical Modelling

In some cases an analytical solution to differential equations describing the phenomenon under study cannot be found in final form. Using various approximate methods (for example, the method of finite differences), such equations can be brought to a form convenient for solving on high-speed digital computers or to a form in which they could be modelled by an ABM. In the latter case we speak of mathematical modelling.

In other words, mathematical modelling represents the solution to differential equations processed by special analogue computers. It follows from such a definition that in order to adopt mathematical modelling it is essential that the phenomenon under study be capable of description by differential equations and that electrical analogues for the parameters describing it can be found. The merits of such computers, compared to digital computers of the discrete type, are low cost, relatively small dimensions and simple operation.

Analogue machines are constructed in two ways: construction of models—analogs directly simulating the process, and construction of models from individual computing devices (cells), for example electric networks, replacing the computational networks in the method of finite differences. Devices of the first type are highly specialised and intended for solving only single or a narrow series of identical problems. Devices of the second type are more universal and capable of solving an entire class of problems (phenomena).

Electronic machines of a mixed type also exist, in which the hardwares of both devices mentioned above are combined. Such machines are used for studying more complex phenomena in which several processes occur under different unique conditions.

The method of electrohydrodynamic analogy (EHDA) was first propounded by the Russian scientist N.I. Pavlovsky as the theoretical basis for mathematical modelling and the very basis for developing electrical machines of the first type (analogue models). This method is built on the analogy between processes of variation in parameters of electrical current and processes of variation in analogous parameters in the fields of mechanical, acoustic, thermal and other physical phenomena. The possibility of replacing approximate methods for solving analytical problems, in particular the method of finite differences, leading to the method of electrical networks (grids) in which the electrical processes are described by equations, similar to those of finite difference equations, and also the possibility of usage of computers, led to development of the second type of machines.

4.1.1 Theory of Analogue Models

Usually by analogy of two physical phenomena of different nature we mean that case wherein the phenomena occur in geometrically similar systems, and corresponding parameters of fields of such phenomena at similar periods of time and in similar points in space have a specific ratio, i.e., under the conditions:

$$t_N/t_M = t_c;\ x_N/x_M = y_N/y_M = z_N/z_M = l_N/l_M = l_c,$$

satisfying the requirements of their proportionality and

$$Q_1/P_1 = Q_2/P_2 = Q_3/P_3 = \ldots = Q_n/P_n = Q_c/P_c,$$

where Q_i and P_i ($i = 1, 2, 3,\ldots,n$) are parameters of fields in the model and the full-scale. In other words, the ratios of corresponding analogous parameters of both phenomena under study should be equal. Here the equations describing the fields would be identical if expressed in a dimensionless manner.

The analogy method is widely used today in scientific investigations. It enables one to transfer very effectively the methods of analysis of phenomena developed in a particular field to uninvestigated new fields. The analogy method may be successfully employed for solving problems using computers or electrical network devices as well as in experimental investigation of processes on analogue machines, by the method of electrohydrodynamic analogues (EHDA). Examples of analogy between processes and individual parameters of physical fields [9] are given below.

Electric field of current: Ohm's law

$$j = -\sigma \text{ grad } U; \ I = \int jds,$$

where j is current density; σ electric conductance; U electric potential; I current.

Potential field of motion: Equation for liquid flow

$$v = - \text{ grad } \varphi, \ Q_n = \int vds,$$

where v is velocity of liquid flow; φ potential of liquid flow; Q_n liquid consumption.

Field of liquid filtration: Darcy's law

$$v = -k \text{ grad } h, \ Q_\varphi = \int vds,$$

where v is rate of filtration; k filtration coefficient; h piezometric head; Q_φ filtrate drain.

Magnetic field: Law of magnetic induction

$$B = -\mu \text{ grad } U, \ \varphi = \int Bds,$$

where B is magnetic induction; U magnetic potential; μ magnetic permeability; φ magnetic flux.

Thermal field: Fourier law

$$g = \lambda \text{ grad } T, \ Q_T = \int gds,$$

where g is heat flow; λ coefficient of thermal conductivity; T temperature; Q_T heat flow.

Let us consider in detail the method of EHDA between the passage of electric current and potential flow of liquid. For simplicity, the flow is considered in the x, y co-ordinates (Fig. 32).

Let us isolate in the electric field and in the liquid, elementary tubes of current with elementary section of length ds. Potentials of electric current and liquid in the sections are denoted by U and $U + \Delta U, \varphi$ and $\varphi + \Delta\varphi$ respectively. Density of electric current and velocity of liquid along the line of current would be correspondingly equal:

$$j = -\sigma dU/ds, \quad v = -d\varphi/ds$$

or in the projections on the axes x and y

$$j_x = -\sigma \partial u/\partial x, \; j_y = -\sigma \partial u/\partial y; \qquad \ldots (4.1)$$

$$v_x = -\partial\varphi/\partial x, v_y = -\partial\varphi/\partial y. \qquad \ldots (4.2)$$

The continuity equation for both fields is of the form:

$$\partial j_x/\partial x + \partial j_y/\partial y = 0; \; \partial v_x/\partial x + \partial v_y/\partial y = 0. \qquad \ldots (4.3)$$

Substituting in equation (4.3) the corresponding values from (4.1) and (4.2), we get the same Laplacian differential equation for both cases, i.e.,

$$\partial^2 U/\partial x^2 + \partial^2 U/\partial y^2 = 0,$$

$$\partial^2\varphi/\partial x^2 + \partial^2\varphi/\partial y = 0. \qquad \ldots (4.4)$$

Consequently, in this case we can model the potential flow of liquid by the passage of electric current in a certain medium, choosing the coefficient of electric conductance in a manner described above and by observing the

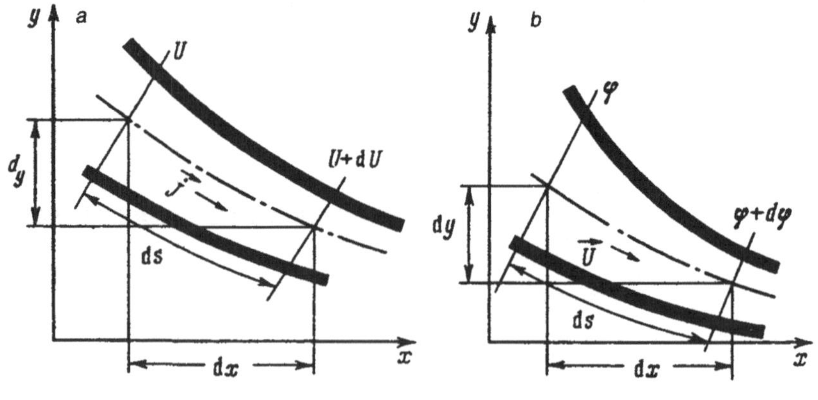

Fig. 32. Illustration of analogy between (a) passage of electric current and (b) potential flow of liquid.

sameness of boundary conditions and condition of geometric similitude, i.e., at the zonal boundaries of the field being studied we should be able to assign in a similar manner either the values of potentials φ and U or the values $\partial\varphi/\partial_n$ and $\partial U/\partial_n$, where n is normal to boundary of the zone. Further, the electric model should represent the zone of liquid flow in the full-scale with a certain scale factor without distortions, while the coefficient of thermal conductivity should be directly proportional to unity (practically to any constant number).

It should be noted here that the analytical solution to Laplacian equation (4.4) for a liquid is complicated. At the same time this equation can be readily solved experimentally for an electric current, as the values of electric potential are easily measured in an experiment. The same equation for a liquid cannot be solved through an experiment as the potential φ does not exist in nature and is introduced conditionally for determining characteristics of potential flow of liquid in pure theoretical investigations.

The method of EHDA is successfully used for studying stationary physical processes described by equations of the elliptical type (Laplace equations with right side or without it). In mining practice, the method of EHDA is used for solving problems of liquid filtration in certain processes and propagation of blast-induced seismic waves.

The EHDA device can serve as a prototype for the analogue model. The principle of operation of such a device is based on adopting the method of bridge (compensation) measurements of electric potential in the model field. The EHDA device (Fig. 33) consists of a feeder block step-down transformer (1), rectifier (2), voltmeter (3), milliammeter (4); measurement block graduated potentiometer (5), zero indicating-galvanometer (6), scanning needle (7), the model itself (8), resistance box (9) and busbar terminals (10).

An electric conducting medium made from materials capable of providing a specific electric conductance of the medium in any section of the model strictly proportional to the value of corresponding parameter in the full-scale, is used as the model. To obtain such results boards and paper conducting electricity are often used, which are capable of electron conduction and permit usage of direct current and provide very low contact resistance between the paper and metallic terminals.

Before fabricating the model, usually the geometric form of the field to be studied, the boundary conditions and the geometric scale of modelling are determined. If the boundary conditions happen to be normal derivatives, equal to zero, of unknown value of voltage, they are easily realised by insulating the contour of the model body, i.e., cutting the paper along edges of the model (creating air insulation). To obtain constant values of voltage at the boundary, busbar terminals are installed at the body contour and the required potential corresponding to the boundary conditions is fed to them.

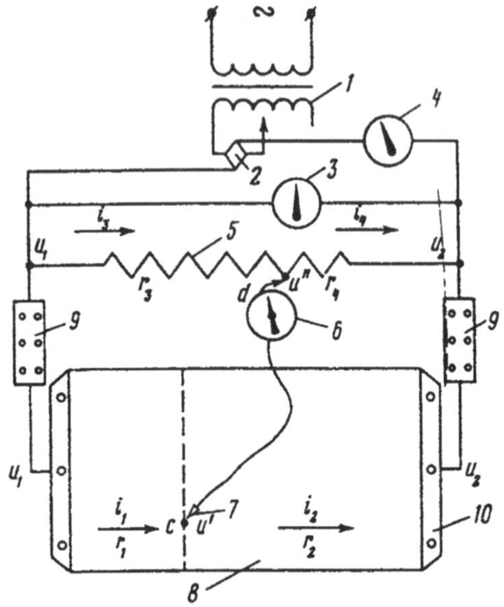

Fig. 33. Highly schematic diagram of the electrohydrodynamic analogue EHDA device. U_1—input voltage; U_2—output voltage; C—Point in the model at which the electric potential U' is equal to the potential of potentiometer U'' at point d.

Methodology of experimentation involves determining the geometric location of points having the same values of electric potential equal to the potential of the potentiometer, i.e., $U' = U''$ and $i = 0$ (see Fig. 33). So-called equipotential lines are obtained by connecting such points with smooth curves. These curves are the analogues for the corresponding parameters in the full-scale.

Lines of current are plotted in a similar manner. While plotting them the property of orthogonality of lines of current and lines of equipotentials should be used and the inverse problem solved, i.e., the assigned boundary conditions should be interchanged, in particular, permeable boundaries should be made impermeable (by removing busbar terminals) and impermeable, permeable (by installing busbar terminals). In other respects, the methodology of plotting lines of current does not differ from the methodology of plotting lines of equipotentials.

4.1.2 Fundamentals of the Theory of Analogue Computers

Let us consider key aspects of the theory of analogue computers through the example of modelling processes which could be described by linear equations as partial derivatives of the second order.

Equation of elliptical type: non-homogeneous (Poisson equation):

$$\frac{\partial}{\partial x}\left(f_1\frac{\partial\varphi}{\partial x}\right)+\frac{\partial}{\partial y}\left(f_2\frac{\partial\varphi}{\partial y}\right)+\frac{\partial}{\partial z}\left(f_3\frac{\partial\varphi}{\partial z}\right)=F(x,y,z);$$

homogeneous (Laplace equation):

$$\frac{\partial^2\varphi}{\partial x^2}+\frac{\partial^2\varphi}{\partial y^2}+\frac{\partial^2\varphi}{\partial z^2}=\nabla^2\varphi=0.$$

These equations characterise steady-state stationary processes of filtration, diffusion, humidification, dewatering etc.

Equations of the parabolic type (equation of thermal conductivity) also belong to this type:

$$\nabla^2\varphi+A\partial\varphi/\partial t=F(x,y,z).$$

This equation characterises the non-steady-state process of heat exchange. At $F(x,y,z)=0$ we have a homogeneous equation of the parabolic type (Fourier equation):

$$\nabla^2\varphi+A\partial\varphi/\partial t=0.$$

Equations of the hyperbolic type also belong to this type:

$$\nabla^2\varphi-B\partial^2\varphi/\partial t^2=F(x,y,z).$$

This equation also characterises non-steady-state processes, in particular undulating vibratory processes.

In these equations φ is the unknown quantity (function of co-ordinates x, y, z and time t), while F, f_i, A, B are the known functions of co-ordinates. To solve such equations, it is necessary to additionally assign initial and boundary conditions. If time is an independent variable, only the initial conditions (Cauchy problem) are assigned. If the co-ordinates happen to be independent variables, the problem is termed as a boundary value problem and is solved only for a specific area by taking into consideration three types of boundary conditions.

1. The unknown function (Dirichlet problem) itself is assigned at the boundary

$$\varphi|_b=\varphi_s,$$

where s is a point on the boundary; [b-denotes boundary—translator].

2. Normal derivative of the unknown function (Neyman problem) is assigned at the boundary:

$$\left.\frac{\partial\varphi}{\partial n}\right|_b=f(s).$$

3. A linear combination of the first two conditions (a mixed problem) is assigned to the boundary.

For non-stationary boundary problems, both boundary and initial conditions are given.

Various approximate methods are used for solving such problems: method of finite differences, method of characteristics, integral method, variation method (Monte Carlo method) etc. The method of finite differences or the network method is widely used. Its essence lies in replacing the derivative by its approximate value expressed as a difference in function values at adjacent points.

For example, let us find the second derivative at a point 0 for a function φ, in the plane x, y defined at points 1, 2, 3 and 4 (Fig. 34, a).

The first derivatives for points I, II, III, IV could be roughly expressed as

$$\left.\frac{\partial\varphi}{\partial y}\right|_{\mathrm{I}} = \lim_{\Delta y\to 0}\frac{\Delta\varphi}{\Delta y} \approx \frac{\varphi_1-\varphi_0}{h_1},$$

$$\left.\frac{\partial\varphi}{\partial x}\right|_{\mathrm{II}} = \lim_{\Delta x\to 0}\frac{\Delta\varphi}{\Delta x} \approx \frac{\varphi_2-\varphi_0}{h_2},$$

$$\left.\frac{\partial\varphi}{\partial y}\right|_{\mathrm{III}} = \lim_{\Delta y\to 0}\frac{\Delta\varphi}{\Delta y} \approx \frac{\varphi_3-\varphi_0}{h_3},$$

$$\left.\frac{\partial\varphi}{\partial x}\right|_{\mathrm{IV}} = \lim_{\Delta x\to 0}\frac{\Delta\varphi}{\Delta x} \approx \frac{\varphi_4-\varphi_0}{h_4}.$$

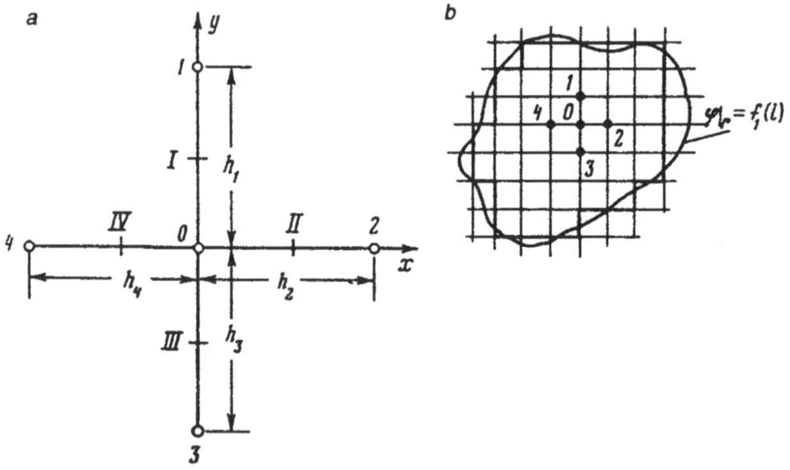

Fig. 34. Scheme for finding the second derivative at junction point 0 according to values of the function (a) at adjacent points; (b) splitting the field with a grid. 1, 2, 3, 4—nodal points; *l*—boundary of field *S*.

Hence, the second derivative at point 0 would be:

$$\left.\frac{\partial^2\varphi}{\partial x^2}\right|_0 = \frac{\left.\frac{\partial\varphi}{\partial x}\right|_{II} - \left.\frac{\partial\varphi}{\partial x}\right|_{IV}}{(h_2+h_4)/2} = \frac{\frac{\varphi_2-\varphi_0}{h_2} - \frac{\varphi_4-\varphi_0}{h_4}}{(h_2+h_4)/2},$$

$$\left.\frac{\partial^2\varphi}{\partial y^2}\right|_0 = \frac{\left.\frac{\partial\varphi}{\partial y}\right|_{I} - \left.\frac{\partial\varphi}{\partial y}\right|_{III}}{(h_1+h_3)/2} = \frac{\frac{\varphi_1-\varphi_0}{h_1} - \frac{\varphi_3-\varphi_0}{h^3}}{(h_1+h_3)/2},$$

and therefore the Laplacian $\nabla\varphi$ would be:

$$\nabla\varphi = \frac{\partial^2\varphi}{\partial x^2} + \frac{\partial^2\varphi}{\partial y^2} = \frac{(\varphi_2-\varphi_0)/h_2 + (\varphi_2-\varphi_0)/h_4}{(h_2+h_4)/2} + \frac{(\varphi_1-\varphi_0)/h_1 + (\varphi_3-\varphi_0)/h_3}{(h_1+h_2)/2} \qquad \ldots (4.5)$$

or, if $h_1 = h_3 = h_y, h_2 = h_4 = h_x$,

$$\partial^2\varphi/\partial x^2 + \partial^2\varphi/\partial y^2 = (\varphi_2 - 2\varphi_0 + \varphi_4)/h_x^2 + (\varphi_1 + 2\varphi_0 + \varphi_3)/h_y^2. \qquad \ldots (4.6)$$

It can be seen from formulae (4.5) and (4.6) that in solving problems using (4.5) the accuracy obtained is of the order h, while that in using (4.6) is of the order h^2; therefore it is better to take the span (step) as equal.

In the case of boundary problem $\nabla\varphi = F(x, y, z)$, for a specific field S, we should assign the function $\varphi(S)$ or $\frac{\partial\varphi}{\partial n}(S)$ along the contour l. Let us now adopt the method of difference for solving the boundary problems.

The field S is divided by a series of mutually perpendicular lines (Fig. 34, b).

The point of intersection of the two lines is termed a node and difference equation (4.5) can be written for any point (inner points) which, after transformation, is written for a homogeneous medium as:

$$\frac{h_1+h_3}{2h_2}(\varphi_2-\varphi_0) + \frac{h_1+h_3}{2h_4}(\varphi_4-\varphi_0) + \frac{h_2+h_4}{2h_1}(\varphi_1-\varphi_0) + \frac{h_2+h_4}{2h_3}(\varphi_3-\varphi_0) = \frac{1}{4}(h_1+h_3)(h_2+h_4)F(x,y). \qquad \ldots (4.7)$$

For a non-homogeneous medium, equation (4.7) can be written as:

$$\frac{h_1+h_3}{2h_2} f_2(\varphi_2-\varphi_0) + \frac{h_1+h_3}{2h_4} f_4(\varphi_4-\varphi_0) + \frac{h_2+h_4}{2h_1} f_1(\varphi_1-\varphi_0) + \frac{h_2+h_4}{2h_3} f_3(\varphi_3-\varphi_0) = \frac{1}{4}(h_1+h_3)(h_2+h_4)\, F(x,y), \qquad \ldots (4.8)$$

where f_1, f_2, f_3, f_4 are known functions of the co-ordinates at points 1, 2, 3, 4 that characterise non-homogeneity of the medium.

Equations of the type (4.7) and (4.8) can be formulated for each inner nodal point. Therefore the number of unknowns in such equations would be equal to the number of inner nodal points, as the values of functions at the boundary are known.

The system of equations thus obtained is solved by the method of sequential approximation. This requires much computational work, necessitating the use of computing and modelling machines.

The method of mathematical modelling is based on the possibility of replacing the method of networks by electric networks in which electric processes are described by equations similar to finite difference equations (4.6) and (4.8). In other words, the differential equations considered above are reduced to finite difference equations which are then modelled by the electric network machines. In this case electric elements are considered as certain analogues of physical quantities. As of today, two systems of electromechanical analogues are being used.

The first system stems from the analogy between the velocity of flow in a mechanical system and electric current in a conductor. This results in the following analogy between mechanical and electrical quantities.

$$p \sim u, \quad v \sim I, \quad H \infty R, \quad m \infty L, \quad R \infty 1/C. \qquad \ldots (4.9)$$

The second system is based on the analogy between equations of velocities in a mechanical system and the balance of voltages in an electric circuit and reduces to analogues of the following type.

$$p \infty I, \quad v \infty U; \quad m \infty C; \quad H \infty 1/R; \quad R \infty L. \qquad \ldots (4.10)$$

In the analogies in (4.9) and (4.10) p, v, H, m, R are pressure, displacement velocity, mechanical resistance, mass and stiffness respectively, while U, I, R, L, C are electric voltage, current, resistance, self-inductance and capacitance respectively.

4.1.3 Computing Devices

Computing devices are modelled according to operations on mathematical equations to be solved. These contain individual operative blocks, each of which fulfils a single mathematical operation: addition, multiplication, integration, differentiation, transformation of functions etc.; in other words, these operative blocks of structural ABM are isomorphic with respect to mathematical operations. In such a case, depending on the type of equation to be solved, a circuit of ABM is created from among the operative blocks in which the mathematical operations of the equation being solved are mapped term-wise. Let us illustrate the circuit of a structural ABM consisting of two integrators (Fig. 35). Each integrator S_1 and S_2 integrates the incoming variable physical quantity with respect to an independent variable (in this case over time) and yields the integral of input physical quantity over time. The

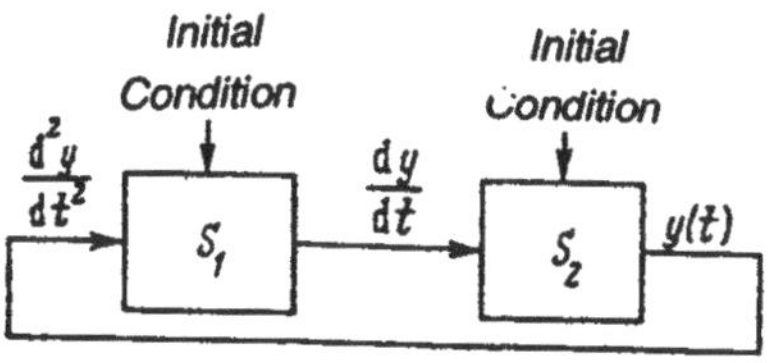

Fig. 35. Circuit of structural analogue computer (ABM) for solving differential equation $d^2y/dt^2 = y$.

S_1—first integrator; S_2—second integrator; $y(t)$—unknown function of time t.

initial value of input variable is assigned prior to integration (this is shown in the diagram of integrator by an incoming arrow with corresponding inscription).

The output variable of integrator 2, $y(t)$, reproduces the solution for the differential equation $d^2y/dt^2 = y$. Actually, if at the outlet of integrator S_2 we have $y(t)$, the derivative dy/dt should be fed at its inlet, which itself is the output quantity from the S_1 integrator. Consequently, the derivative $d/dt(dy/dt) = d^2y/dt^2$ should be fed as input of the integrator S_1, which in accordance with the scheme is equal to the function itself, i.e., equation $d^2y/dt^2 = y$ is fulfilled.

It is to be noted that for solving mathematical problems describing behaviour of a certain physical phenomenon (quantity) on a structured ABM, a very simple circuit for connecting the operative blocks should be chosen so as to effectively represent the entire behaviour of the initial physical system. However, in most cases there is a possibility of not reflecting the direct behaviour of each or certain elements in such a system.

The process of analysing the initial mathematical data of the problem and identifying the method of connecting individual operational blocks of ABM, i.e., programming the ABM, are usually undertaken before solving problems on a structured ABM. While doing so, two key problems are to be tackled:

—Organising the set of ABM blocks into a computational circuit and establishing links between the blocks for transmitting results of computations in blocks from the output terminal of one block to the input of another;

—Representation of initial mathematical variables through physical quantities by choosing corresponding scales.

As a result of programming, we obtain a structured circuit in which the operational blocks are shown by conditional notations while the links between the blocks are indicated by lines.

Solving the problem by modelling on ABM reduces to measuring the values of physical quantities either by measuring devices or recording them on a special oscilloscope which is akin to getting data from experiments.

More convenient and reliable ABMs are electronic and capable of rendering solutions very fast; they have a highly developed electronic instrumentation for measurement, recording and visual observation. Here the electrical voltage of direct current happens to be the physical quantity representing mathematical variables of the problem being solved to a certain scale. Hence the electrical voltage in an ABM is also termed a machine variable.

Blocks in an electronic ABM are connected by electric cables, along which the voltage is transmitted from output terminals of one set of blocks to inlet terminals of another.

4.2 Design of Electric Network Machines and Computing Devices

4.2.1 Electric Network Machines

Let us study the design of electric network machines for modelling stationary and non-stationary physical fields described by equations of elliptical and hyperbolic types.

Network equations for stationary processes are usually of the form represented by relationship (4.8). For modelling this equation, electric circuits are made, representing the same networks into which the area under study was split as in the method of finite differences (Fig. 36).

For each nodal point of the electric network, one of the Kirchoff equations can be written

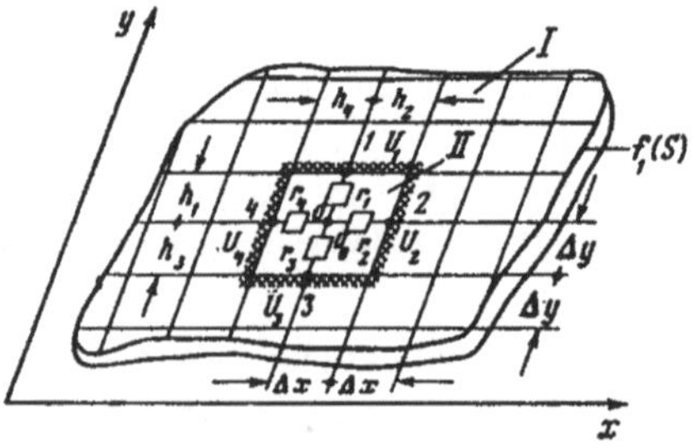

Fig. 36. Replacing a continuous medium with electric circuits.
I—continuous medium; II—electric circuits; 0, 1, 2, 3, 4—nodal points.

$$\sum_{i=1}^{n} I_i - I_0 = 0, \quad \sum_{i=1}^{n} I_i = 0, \qquad \dots (4.12)^*$$

or

$$(U_1 - U_0)/r_1 + (U_2 - U_0)/r_2 + (U_3 - U_0)/r_3 + (U_4 - U_0)/r_4 = 0, \qquad \dots (4.13)$$

$$(U_1 - U_0)/r_1 + (U_2 - U_0)/r_2 + (U_3 - U_0)/r_3 + (U_4 - U_0)/r_4 - (U - U_0)/R = 0, \qquad \dots (4.14)$$

where U_i is the potential at the nodal points of the electric field; r_i the active resistances between nodal points; I_i current; U voltage supplied to the nodal point through active resistance R; I_0 total current at the nodal point.

Assuming that the voltage U is proportional to unknown function φ, i.e.,

$$U_i = k_1 \varphi_i, \qquad \dots (4.15)$$

(where k_1—proportionality coefficient), equation (4.13) can be rewritten as

$$\sum_{i=1}^{n} \frac{U_i - U_0}{r_i} = k_i \sum_{i=1}^{n} \frac{\varphi_i - \varphi_0}{\tilde{r}_i}.$$

Equations (4.8) and (4.14) are similar and could be interchanged if r_i were proportional to the coefficients at finite differences, i.e.,

$$r_i = k_2 \tilde{r}_i, \qquad \dots (4.16)$$

where

$$\tilde{r}_1 = \frac{2h_1}{f_1(h_2 + h_4)}; \quad \tilde{r}_2 = \frac{2h_2}{f_2(h_1 + h_3)};$$

$$\tilde{r}_3 = \frac{2h_3}{2f_3(h_2 + h_4)}; \quad \tilde{r}_4 = \frac{2h_4}{f_4(h_1 + h_3)}; \qquad \dots (4.17)$$

k_2—proportionality coefficient.

If resistances r_i computed according to formulae (4.16) and (4.17) are introduced between nodal points of the electric network, while at the boundary nodal points voltages proportional to values of unknown function at the boundary are supplied, then for determining the function φ_i it is essential to measure voltages at the inner nodal points and their values should be recalculated according to formula (4.15).

Usually the proportionality coefficients k_1 and k_2 are chosen based on the practical aspects of electric machines. Thus, assuming the value of voltage supplied to the machine and maximum resistance of the machine as 100%, the values of k_1 and k_2 are determined from the condition that the maximal values of voltages at the nodal points and at the boundary as well as the maximum computed resistances should be less than the permissible values, i.e.,

$$k_1 \leqslant 100/U_{max}, \qquad k_2 \leqslant 100/r_{max}. \qquad \dots (4.18).$$

---

* Misnumbering in Russian original should read (4.11), or perhaps equation (4.11) omitted in Russian publication—General Editor.

While solving a non-homogeneous differential equation, it should be borne in mind that the right side of equation (4.8) is approximated in accordance with (4.12), to current I_0. This current should be assigned in such a way that it would be proportional to the right side of equation (4.8), i.e.,

$$I_0 = k_3 \frac{1}{4}(h_1 + h_3)(h_2 + h_4)\ F(x, y),$$

where k_3—proportionality coefficient.

According to expression (4.13) $I_0 = (U - U_0)/R$. At $U \gg U_0$

$$I_0 \approx U/R,$$

where U is the voltage supplied to the nodal point through active resistance R.

Therefore, at very low values of voltage in the network, the current may be considered proportional to the value of voltage fed at resistance R. In order to ensure that $|U| \gg |U_0|$, resistance R should be very large vis-à-vis the resistance between nodal points, i.e., $|R| \gg |r_i|$.

The proportionality coefficients k_1, k_2, k_3 are related to each other in the following manner

$$k_1/k_2 = k_3 \text{ or } k_1/(k_2 k_3) = 1. \qquad \ldots (4.19)$$

Consequently, values of coefficients k_1 and k_2 are determined from condition (4.18) and that of k_3 from condition (4.19). The value of R is taken from the catalogue of the machine.

Thus for solving a non-homogeneous problem, it is necessary to formulate an electric network consisting of resistances, as determined by conditions (4.16) and (4.17) and at each nodal point supply through resistance R, voltage proportional to the right side of equation (4.8) at the given point. Further, it is necessary to assign boundary conditions, i.e., at the boundary points in the network voltages, proportional to values of variable φ at the corresponding points in the boundary of area under study, are to be supplied. (Fig. 37).

While drawing electric networks for non-stationary fields, the area under investigation is also divided into a grid and the differential equation replaced by a system of grid equations formulated for each nodal point.

Let us consider the mathematical modelling of equations of the parabolic type; for example, let us find the solution to the Fourier equation in the area S for

$$\frac{\partial}{\partial x}\left[f(x, y)\frac{\partial \varphi}{\partial x}\right] + \frac{\partial}{\partial y}\left[f(x, y)\frac{\partial \varphi}{\partial y}\right] = A(x, y)\frac{\partial \varphi}{\partial t} \qquad \ldots (4.20)$$

at the given boundary conditions:

—initial conditions

$$\varphi(x, y, 0) = f_1(x, y);$$

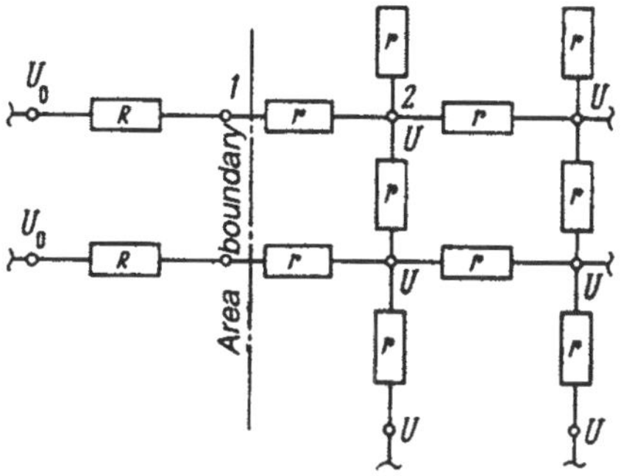

Fig. 37. Assigning the boundary values—schematic.
1—point beyond the boundaries of area under investigation; 2—point within the area.

—boundary conditions

$$\varphi(0, t) = f_2(t);\ \varphi(x_s, y_s, t) = f_3(l, t),$$

where l is the point at the areal boundary.

For equations (4.20) the system of grid algebraic equations would be

$$\frac{h_1 + h_3}{2h_2} f_2(\varphi_2 - \varphi_0) + \frac{h_1 + h_3}{2h_4} f_4(\varphi_4 - \varphi_0) + \frac{h_2 + h_4}{2h_1} f_1(\varphi_1 - \varphi_0) +$$

$$\frac{h_2 + h_4}{2h_3} f_3(\varphi_3 - \varphi_0) = \frac{1}{4}(h_1 + h_3)(h_2 + h_4)A(x, y)\frac{\partial \varphi}{\partial t}. \qquad \ldots (4.21)$$

The left side of equality (4.21) is approximated in a similar manner as adopted in solving stationary problems. The right side is left in the differential form itself. For approximation, current proportional to the first derivative of unknown function with respect to time should be supplied to each nodal point. In order to do so, the capacitor should be connected to the nodal point instead of resistance (Fig. 38).

The model should work on dc (if ac is supplied, the capacitor would not be a differentiating element but act as a resistance, causing only a shift in the phase).

Then, current passing through the capacitor is given by

$$I = CdU/d\tau,$$

where C is the capacitance (of the capacitor); U the voltage on the plates of the capacitor; τ the duration of current flow.

In this case the Kirchoff equation for each nodal point would be of the form;

$$i_1 + i_2 + i_3 + i_4 = I_0,$$

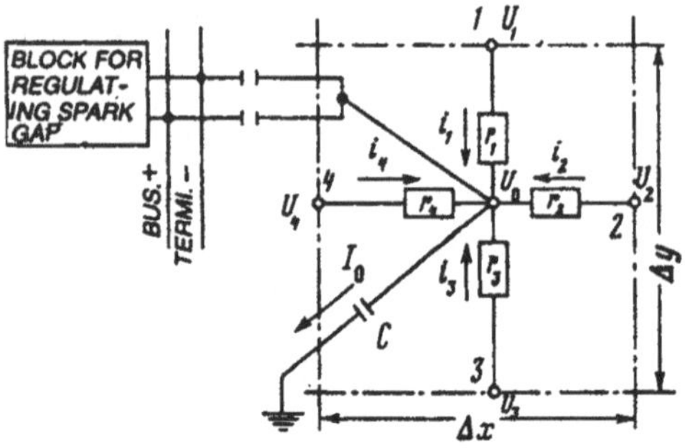

Fig. 38. Circuit for approximating non-stationary physical fields.

or $(U_1 - U_0)/r_1 + (U_2 - U_0)/r_2 + (U_3 - U_0)/r_3 + (U_4 - U_0)/r_4 = CdU/d\tau$.

Introducing the proportionality condition between the parameters of the process being modelled and the electric field,

$$U_i = k_1\varphi_i,\ r_i = k_2\tilde{r}_i,\ \tau = k_4 t;$$
$$C = k_3\frac{1}{4}(h_1 + h_3)(h_2 + h_4)A(x,y)$$

and substituting these in equation (4.21), we obtain

$$\left(\frac{k_1}{k_2}\right)\left[\frac{h_2 + h_4}{2h_1}f_1(\varphi_1 - \varphi_0) + \frac{h_1 + h_3}{2h_2}f_2(\varphi_2 - \varphi_0) + \frac{h_2 + h_4}{2h_3}f_3(\varphi_3 - \varphi_0) + \right.$$
$$\left.\frac{h_1 + h_3}{2h_4}f_4(\varphi_4 - \varphi_0)\right] = \frac{k_1 k_3}{k_4}\left[\frac{1}{4}(h_1 + h_3)(h_2 + h_4)A(x,y)\frac{\partial\varphi}{\partial t}\right]. \quad \ldots (4.22)$$

For interchangeability of equations (4.21) and (4.22), it is essential that the expression for coefficients in the right and left sides of (4.22) be equal to each other, i.e.,

$$k_1/k_2 = k_1 k_3/k_4 \text{ or } k_2 k_3/k_4 = 1. \quad \ldots (4.23)$$

Coefficients k_1 and k_2 are determined from equations (4.15) and (4.16), and coefficients k_3 from the condition $k_3 \leqslant C_{max}/b$, where C_{max} is the maximum capacitance of the capacitors used in the electrical machine; $b = \frac{1}{4}(h_1+h_3)(h_2+h_4)A(x,y)$. Coefficient k_4 is determined from condition (4.23).

A special device—a voltage divider capable of supplying the required fraction of voltage proportional to an unknown function assumed conditionally as unity—is used for assigning boundary conditions while solving the first boundary problem. In such a case, the assigned boundary values of functions are recalculated in percentages and from the divider, also graduated in percentages, the required voltages are fed at contour point.

Boundary conditions assigned in the form of normal derivatives of unknown function are also represented in a finite-difference form, i.e., a normal derivative is replaced by a finite difference of values of the function at two adjacent points abutting the boundary (see Fig. 36):

$$\frac{\varphi_{un} - \varphi_N}{h} \approx \frac{\partial \varphi}{\partial n}. \qquad \ldots (4.24)$$

According to the diagram (see Fig. 37) the current between points 1 and 2 is:

$$I_{1;2} = (U_1 - U_2)/r = (U_0 - U_1)/R.$$

Taking $R \gg r$ and neglecting U_1/R, we get

$$I_{1;2} = U_0/R.$$

Finally, assigning U_0 proportional to the right side of relationship (4.24), we get the necessary relationship between U_1 and U_2, i.e., between φ_{un} and φ_N respectively.

Initial conditions for the function as well as the normal derivative at the corresponding boundary at time $t = 0$ are assigned also by the voltage divider at each point through a relay which, at $t = 0$, feeds the required voltages to the grid and immediately cuts them off.

Solving network (grid) equations by electric analogue machines finally boils down to measurement of voltages and currents in the model under study. The following methods of measurement are used in the process:

Direct measurement using measuring device: accuracy of the method is limited by accuracy of the device (not more than 2–3 places); to prevent distortions the device itself should be of high resistance.

Substitution method: sequential comparison with a certain known quantity usually from the oscillograph (simple, transparent and good accuracy).

Null (compensation) method: similar to the methods of substitution but a zero-indicator is used for comparison, which fixes nil difference between the known and unknown quantities (this method is widely adopted).

Differential (difference) method: here instead of an unknown quantity the difference between it and the known quantity is measured with the help of the above methods (the compensation method is very often used).

Mathematical analogue computing machines of a special type are used for solving the above differential equations in partial derivative form of the Laplace, Poisson and Fourier type. The display type electric grid indicators EI-12 and universal network models UCM-1 are widely used.

4.2.2 Structured Analogue Computers

Apart from modelling through electric networks, where operation of the modelling device is based on direct correspondence between formulation of the finite-difference method and construction of the electric network (see Fig. 36), i.e., based on direct correspondence between elements of the prototype and electric elements of the model (resistances, capacitors etc.), it is also possible to model on electronic ABMs in which operational blocks perform the mathematical operations for solving problems. These are intended for solving systems of general differential equations in partial derivative form. In addition, they can do other mathematical operations, such as addition, subtraction, multiplication, integration as well as generation of differential functions [14].

Structural analogue computers consist of the following main parts: operational blocks intended for mathematical operations; field kits (commutation) for linking various operational blocks and for the formation of structured circuits for solving mathematical problems; controlling devices for ensuring time-dependent interaction of all parts of the ABM, blocks for working the ABM in one of the four regimes (preparation, solving the problem, automatic interruption for feeding intermediate values and changing the commutation between blocks) and for bringing the computer back to the initial state; measuring and recording instrumentation (ranging from simple pointer devices to various oscillographs with recorders for fixing the change in values of variable quantities over time); feeders for rectifying voltages of commercial frequency into the electric voltage ratings essential for the functioning of all parts in the ABM. These voltages should be highly stable, as any variation in such voltage ratings while solving problems on the ABM, causes variation in the machine variable.

The operational blocks are the most important part in an ABM. These can be divided into the following major blocks: summation (summators), integration (integrators) and differentiation, multiplication, non-linear functions (functional transforms) and other auxiliary blocks (blocks for delay, programmed control).

Most operational blocks contain an operational amplifier (OA), schematically shown in Fig. 39, a. In this figure, e_g is input voltage; U_{out} the output voltage and k the voltage amplification factor of OA. The operational amplifier possesses certain properties which are to be kept in mind while working on an ABM.

The amplification coefficient of OA is very high, for very good quality ABMs $k \geq 4 \times 10^4$ to 10^8.

Input and output voltages of OA have different signs (inverse) due to the odd number of cascades (usually three).

The output voltage should vary within the limits of E_{max} of the ABM, i.e., $-E_{max} \leqslant U_{out} \leqslant E_{max}$, otherwise non-linear distortions arise in OA. In

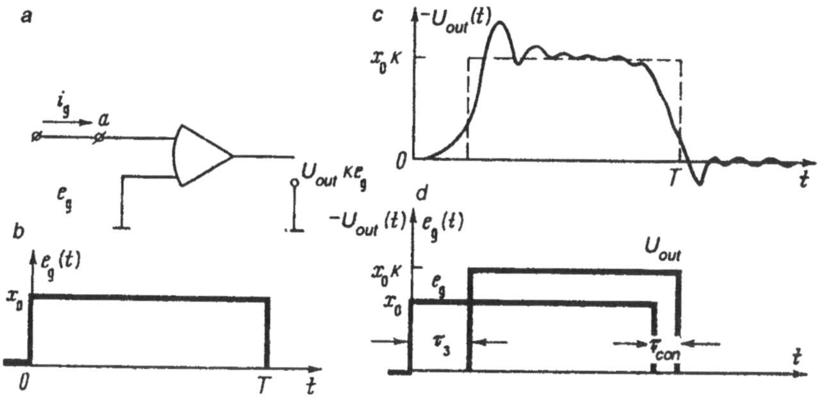

Fig. 39. Schematic diagram of an operational amplifier (OY) and its characteristics. a—electric circuit of OA; b, c—diagrams of rectangular input signal incoming to OA and output signal from OA respectively; d—diagram showing delay τ_z and consequence of input signal τ_{con} on OA.

the lamp-type ABMs usually $E_{max} = 100$ V and in the semi-conductor type ABM, $E_{max} = 10$ to 50 V.

The incoming terminal OA (point *a*) is potentially earthed, i.e., voltage at the inlet of OA is very small and coincides with the potential of common point in the circuit of OA termed 'earth'. Actually, $U_{out} = -e_{-g}k$, hence $e_g = -U_{out}/k$, and if $U_{out} = 100$ V and $k = 4 \times 10^4$, then voltage at the inlet of amplifier $e_g = 2.5$ mV.

In the OA, inlet current i_g is practically absent, as $i_g = e_g/R_{in}$, where e_g is less and R_{in} for the OA is very high (R_{in} is determined by insulation between input contacts of OA).

The operational amplifier possesses certain inertia (see Fig. 39, b, c), i.e., delaying feature and its consequence, schematically shown in Fig. 39, d. The higher the rate of variation in the input voltage e_g, the greater the inertia.

Most OAs possess zero-drift voltage due to variation in emission capabilities of lamps, variation in voltage ratings of power sources and fluctuations in humidity and temperature of the surrounding medium. Hence while solving problems on the ABM, it is periodically necessary to verify the deviation of zeroes in the amplifiers and to restore them to the correct position by means of special devices.

The summation block is shown in Fig. 40, a. It sums up the incoming voltages, simultaneously multiplying them with certain constant coefficients.

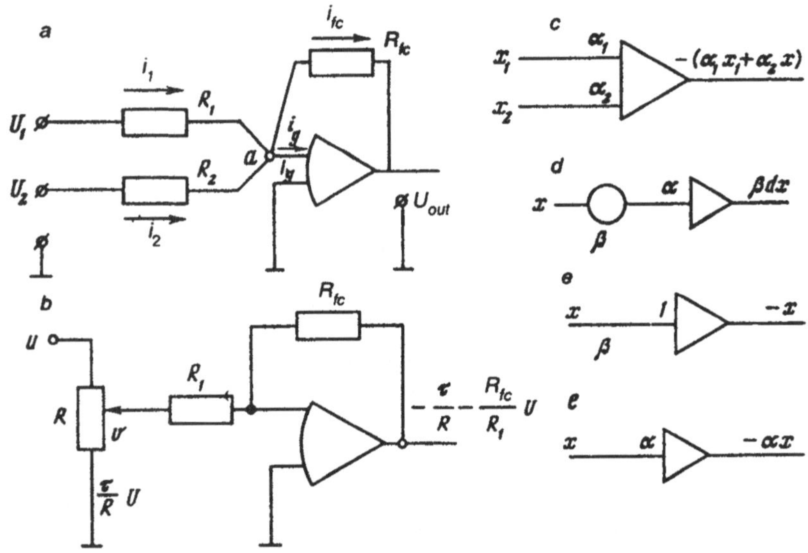

Fig. 40. Simple electric circuits for the operational blocks of ABM.
a—summation block of input voltages; b—summation block with variable coefficient of transmission with potentiometer R; c—structural mapping of summator with transmission coefficients α_1 and α_2 as indicated at the entry to summator; d—structural mapping of summator with varying transmission coefficient β; e—invertor effecting change only in the sign of incoming quantity; f—block for changing input quantity to a constant transmission coefficient with inversion of sign.

Actually, in this circuit two inlets and one outlet are shown. Let us indicate the incoming voltages by U_1 and U_2, which through resistors R_1 and R_2 are fed to the inlet of OA. The operational amplifier is covered by a negative feed back with the help of a resistor R_{fc}, which may be both negative (reducing the input quantity) or positive (increasing the input quantity). In the circuit being studied, this feedback is negative due to the inverting property of OA.

According to the Kirchoff law, for node *a* we have the equation of currents

$$i_{fc} + i_g = i_1 + i_2, \qquad \dots (4.25)$$

where i_{fc} is the current flowing from node *a* through resistor R_{fc}; $i_g = 0$ the current flowing from node *a* to the inlet of OA; i_1, i_2 are currents flowing through resistors R_1 and R_2 respectively.

According to Ohm's law, equation (4.25) is written as

$$(e_g - U_{out})/R_{fc} = (U_1 - e_g)/R_1 + (U_2 - e_g)/R_2.$$

Considering voltage e_g at the inlet of amplifier OA, $e_g \approx 0$, we have

$$U_{out} = \frac{R_{fc}}{R_0}U_1 - \frac{R_{fc}}{R_2}U_2 = -(\alpha_1 U_1 + \alpha_2 U_2),$$

where $\alpha_1 = R_{fc}/R_1$ and $\alpha_2 = R_{fc}/R_2$ are coefficients of transmission at inlets 1 and 2 respectively. These are dimensionless and expressed as a ratio of resistances. To obtain the variable coefficient of transmission at the inlet of the summator, a potentiometer is installed at the inlet (Fig. 40, b) in which impedance R and arm (branch) resistance r are such that voltage at the outlet of the amplifier is

$$U_{out} = -rR_{fc}U_1/(RR_1) = -\beta\alpha U_1,$$

where $\beta = r/R$ is the transmission coefficient of the potentiometer, $0 \leqslant \beta \leqslant 1$. In this case the common coefficient of transmission would vary in the interval $0 \leqslant \alpha\beta \leqslant \alpha$.

In structural diagrams the summator is shown in the form of triangles (Fig. 40, c) indicating the transmission coefficient at each inlet, while the potentiometer is shown as a circle, around which the value of transmission coefficient β is shown (Fig. 40, d).

If we have one inlet and $\alpha = 1$, the summator is termed an inverter, which serves the purpose of reducing a subtraction operation to a summation operation. With one inlet and $\alpha \neq 1$, the summator is termed a scale amplifier $(U_{out} = -\alpha U_{in})$.

An integration block (integrator) is shown in Fig. 41, a. It consists of an OA covered with feedback through a capacitor of capacitance C, with an inlet resistor R. In this case the Kirchoff law for nodal point a is written as

$$(U_{in} - e_g)/R = C\frac{d}{dt}(e_g - U_{out}),$$

since $i = dQ/dt$ (where Q—current at plates of the capacitor, $Q = CU_c$), and $U_c = e_g - U_{out}$. Assuming $e_g = -U_{out}/k = 0$; a differential equation with respect to $U_{out}(t)$ is obtained:

$$dU_{out}(t)/dt = -U_{in}(t)/(RC) = -mU_{in}(t). \qquad \ldots (4.26)$$

Considering that the output quantity in the circuit shown in Fig. 41, is $U_{out}(t)$, it is better to write equation (4.26) in an equivalent integral form

$$U_{out}(t) = -m\int_0^t U_{in}(t)dt + U_0,$$

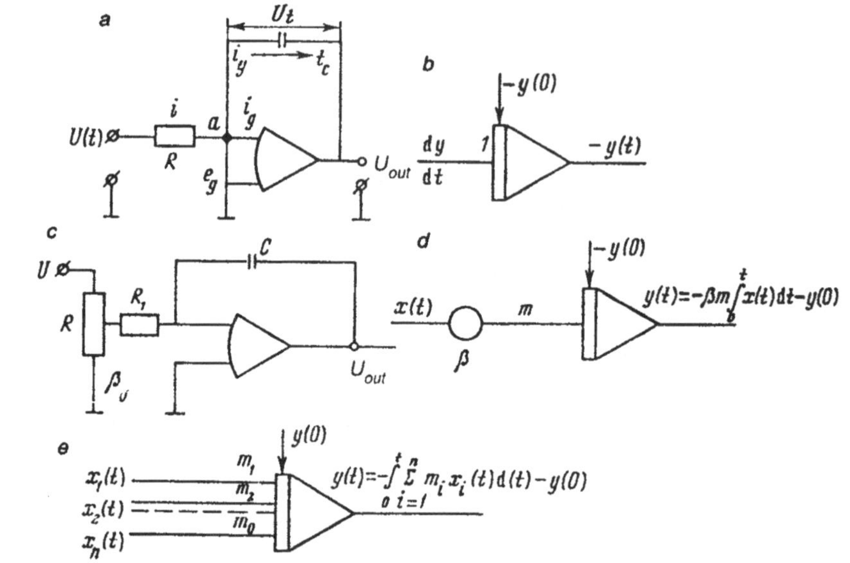

Fig. 41. Electric and structural circuits of integrating operational blocks.
a, b—electric and structural circuits respectively for integrating single variable over time; c, d—electric and structural circuits of integrator with varying transmission coefficient or varying duration of integration; e—structural diagram of an integrator-summator.

where U_0 is the integration constant, equal to U_{out} at the initial moment, i.e., $U_0 = U_{out}(0)$. This indicates that the capacitor is charged to voltage U_0 prior to integration.

The quantity $m = 1/(RC)$ is termed the transmission coefficient of the integrator, with the dimension s^{-1}, which signifies its relationship with the duration of integration.

In structural circuits, integrators are indicated as shown in Fig. 41, b. While working on an ABM, one should be able to assign the transmission coefficients for integrators and initial conditions of integration. The operator should also know the specific working features of the integrator under different regimes: regime of introducing coefficients; regime of assigning initial conditions; regime of integration and regime of functions related to the solution.

It is to be noted that just as in summators, potentiometers capable of giving any value of transmission coefficients βm, where $\beta = r/R$, can be plugged in at the inlet of the integrator. The electric and structural circuits of the integrator with potentiometer are shown in Fig. 41 c, d. The block of integral-summator can also be used if at the inlet of the integrator several voltages are plugged in (Fig. 41, e), for example n inlets.

In such a case the output voltage is given by

$$U_{out} = -\sum_{i=1}^{n} m_i \int_0^t U_i(t)dt + U_0.$$

The differentiation block is shown in Fig. 42, a. In accordance with the electrical layout depicted, the Kirchoff equation for node *a* would be $i_c = i_R$, which can be expressed in terms of voltage as

$$C\frac{d}{dt}(U_{in} - e_g) = (e_g - U_{in})/R,$$

and taking into account $e_g = U_{out}/k = 0$, we have

$$U_{out}(t) = -RC\frac{dU_{in}}{dt}.$$

However, the circuit as considered above is rarely used in an ABM, as it amplifies disturbances (noise). So the operation of differentiation is either not used in an ABM or combined differential blocks are employed, which in addition to differentiation, concomitantly perform smoothening. Two such

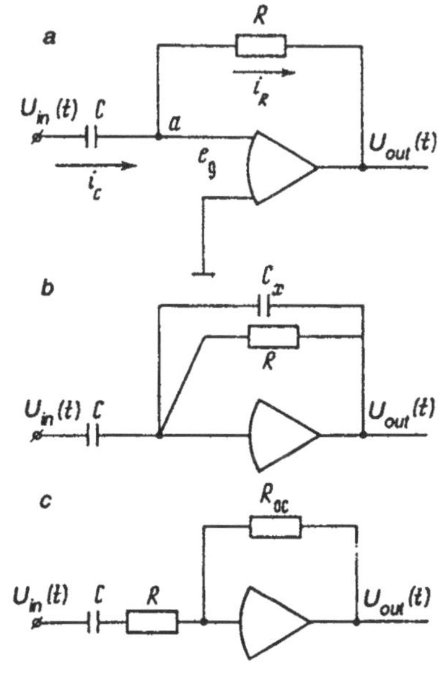

Fig. 42. Electric circuits for blocks of differentiation.
a—noise-wise unstable circuit for differentiating the variable $u(t)$; b, c—combined schemes of differentiating blocks with smoothening of noises.

circuits are shown in Fig. 42, b. In the first circuit the effect of smoothening is achieved by introducing an additional capacitor in the feedback circuit; in the second, by adding an additional resistor in the incoming side.

While solving a series of problems pertaining to the fields of automatic control and regulation, stabilisation theory, wave processes etc. on an ABM, it is necessary to reproduce diverse and complicated transfer functions by means of the operational amplifier [13]. Let us consider this through an example of a linear differential equation between the variables: $x(t)$ and $y(t)$ of the type

$$d^2x/dt^2 + 2ndx/dt + \omega_2 x = ay + bdy/dt, \qquad \ldots (4.27)$$

which in an operator form (for $P = d/dt$) is written as

$$P^2x + 2nPx + \omega^2 x = ay + bPy \qquad \ldots (4.28)$$

or $$(P^2 + 2nP + \omega^2)x(t) = (a + bP)y(t). \qquad \ldots (4.29)$$

If each of the terms in (4.28) are transformed into Laplacian, we can get an equation for mapping the variables x and y in a complex domain, from which an expression for transfer function can be readily determined:

$$x(P)/y(P) = (a + bP)/(P^2 + 2nP + \omega^2). \qquad \ldots (4.30)$$

This function is similar to the symbolic operator form of writing the initial differential equation (4.27) obtained from (4.29), i.e.,

$$x(t)/y(t) = (a + bP)/(P^2 + 2nP + \omega^2).$$

It follows that if on the operational amplifier we can realise expression (4.30) by choosing resistances $Z_{in}(P)$ and $Z_{fc}(P)$, then the variable at the exit $x(t)$ would be the solution to equation (4.28) at null initial conditions and under the given input effect $y(t)$. The transmission coefficient of the problem-solving amplifier is given by

$$\frac{U_{out}(P)}{U_{in}(P)} = -\frac{Z_{fc}(P)}{Z_{in}(P)} = -\frac{y_{in}(P)}{y_{fc}(P)}, \qquad \ldots (4.31)$$

where $y_{in}(P)$ and $y_{fc}(P)$ are conductance of incoming circuit and feedback circuit respectively.

While deriving expressions for transfer functions for the given circuits, one should use either operator expression for total resistance Z_{in} or Z_{fc} when resistors and capacitances are connected in a series, or operator expression for conductance y_{in} or y_{fc} when resistors and capacitors are connected in parallel. For more complex circuits, equations for currents and voltages are derived for all sections in the circuit and these are solved with respect to voltages U_{out} and U_{in}. Two circuits are illustrated in Fig. 43.

For the scheme in Fig. 43, a, we have:

$$Z_{in} = R_1 + \frac{1}{CP}; \; Z_{fc} = R_2; \; Z_{in} = (R_1CP + 1)/(CP).$$

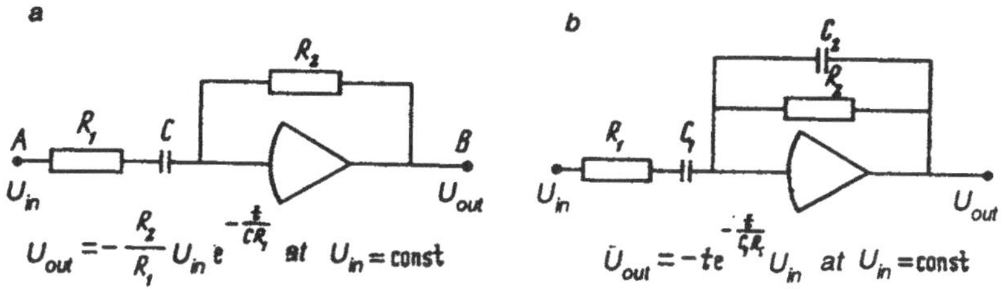

Fig. 43. Electric circuits for blocks of function converters.
a—transforming input function into exponential; b—transforming input function into a more complex exponential function.

Substituting Z_{fc} and Z_{in} in (4.31), the transfer function is obtained

$$\frac{U_{out}(P)}{U_{in}(P)} = -\frac{R_2CP}{R_1CP+1}.$$

For the circuit in Fig. 43, b, we have:

$$Z_{in} = \frac{R_1C_1P+1}{C_1P}; \quad \frac{1}{Z_{fc}} = \frac{1}{R_2} + C_2P = \frac{1+R_2C_2P}{R_2}.$$

The transfer function is given by

$$\frac{U_{out}(P)}{U_{in}(P)} = -\frac{Z_{fc}}{Z_{in}} = -\frac{R_2C_1P}{(1+R_1C_1P)(1+R_2C_2P)}.$$

It should be noted that at $U_{in} =$ const, both circuits give the schemes for generation of exponential functions of a variable (time) t of these types:

for the scheme in Fig. 43, a

$$U_{out} = -R_2U_1e^{-t/CR_1}/R_1.$$

for the scheme in Fig. 43, b

$$U_{out} = -tU_1e^{-\frac{t}{C_1R_1}}$$

Apart from the linear blocks discussed above, non-linear operational blocks or functional converters are also used in an ABM: blocks of non-linear function of one or several variables, blocks for multiplication and division etc. The blocks of non-linear function are divided into universal and specialised. The first group enables one to generate a wide class of functions with the help of a single device by varying its parameters. The second type of blocks is designed for generating only one relationship (exponential, trigonometric etc.). Their construction is based on diodes, non-linear resistors, functional potentiometers and electron-beam tubes.

Most of the ABM-based functional converters realise the function of only a single variable, $y = f(x)$, due to practical difficulties in constructing converters of multiple variables. These are indicated in the structural schemes,

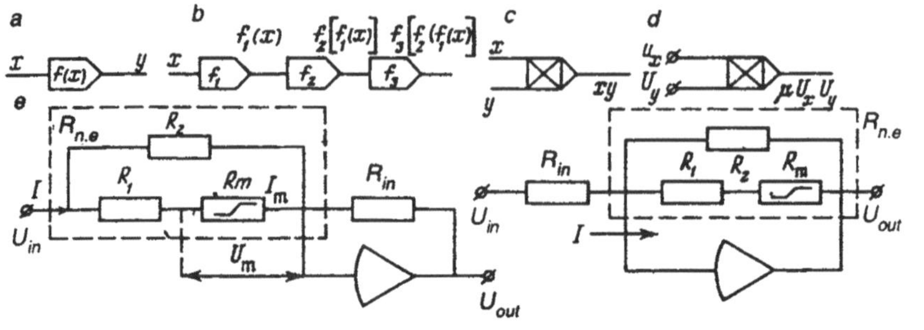

Fig. 44. Circuits of (a, b) functional converters; (c, d) multiplication blocks; (e) blocks for generating squares and square root.

as shown in Fig. 44, a. If several converters are available in an ABM, it is possible to generate more complex functions by connecting them in different ways. Thus in the most simple case, by connecting these in a series, it is possible to superimpose the given function (Fig. 44, b).

The blocks of multiplication and division are intended for realising the function of two variables of the type

$$z = k_1 xy; \ z = k_2 x/y.$$

Theoretically, the multiplication block can be represented as a functional converter of two input variables (a direct acting device). In practice this is difficult to implement and usually for creating blocks of multiplication another (indirect) approach is adopted. Thus in many of the medium-sized and small ABMs, multiplication blocks are structured in accordance with one of the two identically equal transforms

$$xy = \frac{1}{4}\left|(x+y)^2 - (x-y)^2\right|,$$

$$xy = \frac{1}{2}\left|(x+y)^2 - x^2 - y^2\right|.$$

In this case, for multiplication operation we need summators and functional converters, i.e., squaring up devices which square the input variable [13]. The multiplication block is indicated in the structural circuits, as shown in Fig. 44, c and 44, d. In the latter case the electrical nature of machine variables U_x and U_y is emphasised. Here, a voltage proportional to the product of voltages and the proportionality coefficient μ-scale of the multiplication block scheme, is generated at the exit.

Functional converters capable of deriving squares and square roots are structured on the basis of non-linear thyrite or willite semi-conductor resistances, possessing characteristics close to quadratic [13]. Thus, the non-linear element $R_{n.e}$ consisting of a circuit in which the thyrite semi-conductor resistance is connected in a series with resistor R_1 while resistor R_2 is parallel, when connected to the incoming circuit of the operational amplifier (Fig. 44, d) performs the operation of squaring, i.e., $U_{out} = -k_1 R_{fc} U_{in}^2$. The same element, $R_{n.e}$, connected to the feedback circuit of the operational amplifier enables derivation of square root (inverse of the previous operation), i.e.,

$$U_{out} = -\sqrt{U_{in}/(k_1 R_{in})} = -k\sqrt{U_{in}}.$$

To achieve this, schemes for inversion of functions can also be used. Here, at the entry of the integrator obtained from the operational amplifier, a block generating the inverse function $y = f(x)$ is connected through feedback. Depending on the sign of the derivative of this function $f'(x)$, we obtain the scheme either shown in Fig. 45, a if $f'(x) > 0$ or that shown in Fig. 45, b, if $f'(x) < 0$ [14]. In the latter case, the negative sign of the derivative in the scheme is realised by activating the invertor in tandem with the block $y = f(x)$ (see Fig. 45, b).

By analogy with these circuits, performing the division operation reduces to the multiplication operation with the help of a multiplier block connected in the feedback with the operational amplifier OA (Fig. 46, a). In this case determination of the quotient $z = x/y$ of the variables x and y amounts to solving the implicit equation $F(z) = zy - x = 0$, relative to z. The operation of finding the square root (Fig. 46, b) also reduces to solving the implicit equation $F(y) = y^2 - x = 0$ relative to y (where $y = \sqrt{x}$).

The structural circuits shown in Fig. 46, a and b enable one to solve two differential equations respectively:

$$z' = -k[zy - x], \qquad z(0) = 0$$
$$y' = -k[y^2 - x], \qquad y(0) = 0.$$

Diode blocks are often used in the ABM to create logical circuits: restricting the limits of variation of a variable by the non-contacting relay scheme; generating modulus of the function; selection of maximum and minimum

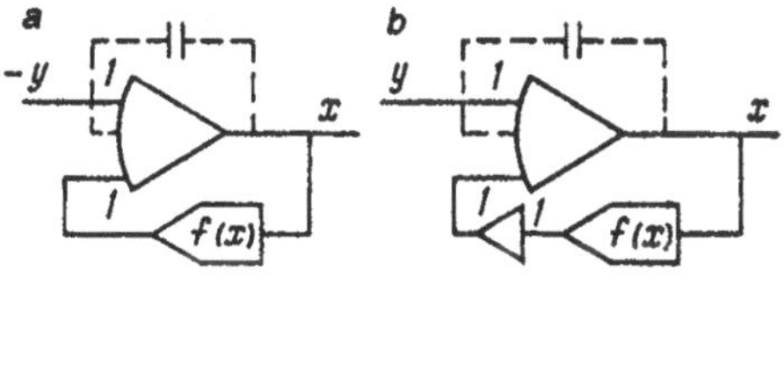

Fig. 45. Electric circuits for solving function $y = f(x)$ depending on sign of the derivative $f'(x)$. a—at $f'(x) > 0$; b—at $f'(x) < 0$.

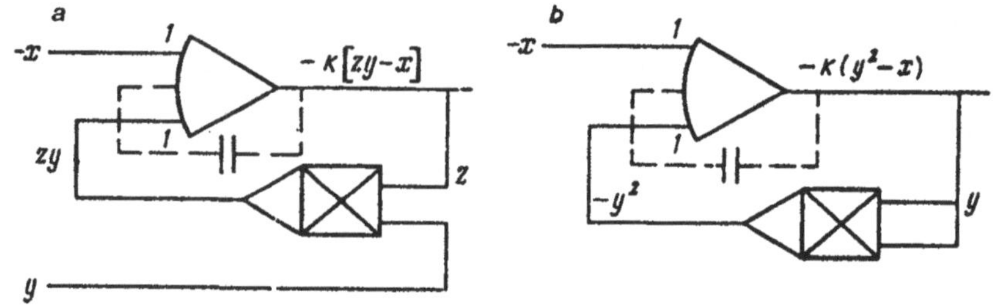

Fig. 46. Circuits for (a) division and (b) finding square root based on inverse functions of initial functions.

value; programmed control and certain other schemes. The property of diodes—to have various resistances depending on the polarisation of applied voltage—is exploited in all these circuits. Let us illustrate the principle of working of diode circuits through the block (scheme) of programmed control shown in Fig. 47, a. This scheme contains an OA which has a negative feedback through diode *D*. A relay coil *P* is connected at the exit of the OA. Contacts of this relay are positioned in the control device of the machine and are also extended to the commutation field of the machine.

In accordance with the circuit in Fig. 47, a, voltage at the exit of OA is determined by the expression

$$U_{out}(t) = -\frac{R_{diode}}{R} U_{in}(t),$$

where R_{diode} is diode resistance, whose value varies depending on whether the diode is open or closed by the voltage applied to it.

For the case shown in Fig. 47, b, as long as the input voltage is positive (up to time t^*) the diode is open and its resistance is low. After time t^*,

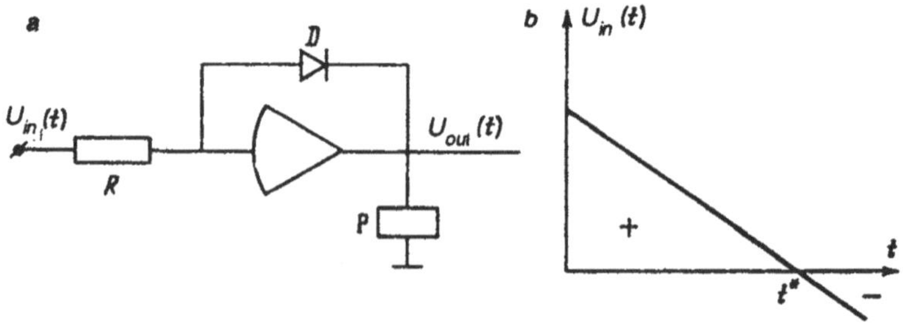

Fig. 47. Programmed control circuit.
a—electric circuit of programmed control block with diode *D* and relay *P*; b—operating scheme of diode passing input voltage up to the moment t^*.

the voltage U_{in} would be negative when the diode closes and its resistance becomes very high. Consequently, at the exit of OA, up to the moment t^* the voltage is practically zero, but at the moment t^* there is a sharp rise in voltage, the value of which is restricted by the resistance of relay coil P, connected to the circuit parallel to the diode, as the entry point of OA is potentially earthed. Thus, at the moment of sharp rise the relay P is activated, opening some contacts and closing others. Here the relay contacts positioned in the control panel of the ABM could be used for automatic shut-off (automatically stopping the solution process). The relay contacts positioned on the field kit of the ABM could be used for automatic variation of the commutation scheme. This is widely used in computational schemes with a variable structure.

4.3 Simulation of Blast Effect by Method of Electro-Hydrodynamic Analogy

The method of electrohydrodynamic analogies (EHDA) is currently used in investigations in various fields of physics, mechanics and hydraulic engineering. This method was applied by O.E. Vlasov and S.A. Smirnov for solving a set of problėms in blasting practice, including those for simulating the fragmentation process of rocks by blasting. The following assumptions were made in applying the theory of modelling:

1. Total quantity of energy Q released in a blast is taken as the key characteristic of blasting.

2. Blast energy is transmitted to the surrounding medium almost instantaneously, while movement of the medium caused by the blast occurs with certain finite speeds, i.e., the process of mechanical effect of a blast is conditionally split into two sequential phases: transmission of blast energy to the medium and movement of the medium itself.

3. At the moment of transmission of blast energy, no real deformations would occur in the medium; in other words, breakages would be absent. This is possible if at this particular moment the medium behaves like an ideal incompressible fluid.

4. Certain finite deformations occur at the movement phase of the medium, which on attaining critical values result in breakage of the medium.

These assumptions greatly simplify the problem and enable final results to be obtained pertaining to the blast effect on a solid medium by means of simple computations and adoption of the electrohydrodynamic analogy.

4.3.1 Fundamentals of Simulating Rock Fragmentation by Blasting, Using the Method of Hydrodynamic Analogy

In conformity with the assumptions made given above, transmission of blast energy in the first phase amounts to determining the state of ideal incompressible fluid under the effect of a specific blast pulse $S = pdt$ (where p—gas pressure), whose value at different points in the medium differs and is a function of co-ordinates. Considering the effect of a blast pulse on an elementary particle in the medium ($dx\ dy\ dz$) and taking into account Newton's second law on pulse effect being equal to quantum of displacement, O.E. Vlasov and S.A. Smirnov obtained the following equations for velocity components of the elementary particle v_x, v_y, v_z through values of blast pulse and density of the medium ρ:

$$v_x = -\frac{\partial}{\partial x}(S/\rho);\ v_y = -\frac{\partial}{\partial y}(S/\rho);\ v_z = -\frac{\partial}{\partial z}(S/\rho). \qquad \ldots (4.32)$$

Whence it follows that velocities arising in the blast have a potential

$$\varphi = S/\rho \qquad \ldots (4.33)$$

whereby expression (4.32) can be written in terms of φ:

$$v_x = -\partial\varphi/\partial x;\ v_y = -\partial\varphi/\partial y;\ v_z = -\partial\varphi/\partial z. \qquad \ldots (4.34)$$

The velocity potential φ in accordance with (4.32) would obviously be equal to zero when $S = 0$, i.e., somewhere at an infinite distance from the blast site or at the free face. It would be constant over the entire surface at the interface charge-medium i.e., $\varphi =$ const. Moreover, the velocity potential according to the assumption concerning incompressibility of the medium at the moment of transmission of blast energy and the resultant condition about continuity in the movement of medium, given by

$$\partial v_x/\partial x + \partial v_y/\partial y + \partial v_z/\partial z = 0,$$

should satisfy Laplace's equation, if values for velocity components as per (4.32) are substituted in this condition,

$$\partial^2\varphi/\partial x^2 + \partial^2\varphi/\partial y^2 + \partial^2\varphi/\partial z^2 = 0.$$

To establish the relationship of potential φ with charge parameters and the medium, we use an idealised concept on the condition of transmission of energy to the medium surrounding the charge, i.e., total kinetic energy of the medium should be equal to the blast energy. This condition results in the relationship

$$\frac{\rho}{2}\int\int\int(v_x^2 + v_y^2 + v_z^2)dx\ dy\ dz = Q. \qquad \ldots (4.35)$$

As the velocities v_x, v_y, v_z possess potential, the computation of volume integral in the left side of (4.35) can be replaced as per Green's formula by computing the surface integral. Then (4.35) is written as

$$\frac{\rho}{2}\oint\varphi\frac{\partial\varphi}{\partial n}\ dF = Q, \qquad \ldots (4.36)$$

where dF is the element of surface limiting the medium under consideration; n the direction of internal normal.

The relationships and conclusions obtained above are applicable in general for charges of any form.

For spherical charge of radius R_{0_z}, due to the central symmetry of movement of the medium, taking into account its incompressibility, the following expressions are derived for the parameters of field of velocities

$$\varphi = \frac{1}{r}\sqrt{\frac{R_{0_z}Q}{2\pi\rho}}; \quad V_r = \frac{1}{r^2}\sqrt{\frac{R_{0_z}Q}{2\pi\rho}}.$$

At the charge surface $\varphi_{surf}|r = R_{0_z}$ we have

$$\varphi_{surf} = \sqrt{Q/(2\pi\rho R_{0_z})}. \qquad \ldots (4.37)$$

For an extended cylindrical charge of finite length, it is necessary to solve the three-dimensional problem in the theory of potentials. In the published literature, a solution to the problem has been given wherein the charge is represented as a triaxial ellipsoid or an ellipsoid of revolution. In the latter case the equipotential surfaces may be represented in cylindrical co-ordinates (putting $r^2 = x^2 + y^2$), as

$$r^2/(a^2+\lambda) + z^2/(b^2+\lambda) = 1,$$

where a is a minor semi-axis (radius of rotation); b a major semi-axis of the ellipsoid; λ a certain parameter varying within the limits $0 \leqslant \lambda \leqslant \infty$ and determining the value of velocity potential φ, i.e., $\varphi = f(\lambda)$.

The velocity potential

$$\varphi = A\int_{\lambda}^{\infty} \frac{d\lambda}{(a^2+\lambda)\sqrt{b^2+\lambda}},$$

where A is a constant determined from the condition of fulfilling the energy balance, i.e., from equation (4.35).

After a set of transformations, the final expression for φ would be

$$\varphi = \frac{2A}{\sqrt{b^2-a^2}} \ln \frac{\sqrt{b^2+\lambda}+\sqrt{b^2-a^2}}{\sqrt{a^2+\lambda}}, \qquad \ldots (4.38)$$

where $$A = \sqrt{\frac{Q\sqrt{b^2-a^2}}{8\pi\rho \ln\left(\dfrac{b+\sqrt{b^2-a^2}}{a}\right)}}. \qquad \ldots (4.39)$$

At the surface of the charge $\lambda = 0$, so the expression for potential (4.28), taking into account the value of A as per (4.39), would be

$$\varphi_{surf} = \sqrt{\frac{Q\ln(b+\sqrt{b^2-a^2})}{2\pi\rho\sqrt{b^2-a^2}}},$$

where φ_{surf} is velocity potential at charge surface; ρ the density of the medium; Q blast energy of the charge.

Usually extended charges with $b \gg a$ are used in blasting operations. This greatly simplifies the expression for φ_{surf},

$$\varphi_{surf} = \sqrt{\frac{Q}{\pi \rho l_z} \ln\left(\frac{2l_z}{d_z}\right)}, \quad \ldots (4.40)$$

where $l_z = 2b$ is the charge length, m; $d_z = 2a$ the charge diameter, m.

The expressions thus obtained for the velocity potential φ_{surf} at the surface of spherical (4.37) and cylindrical (4.40) charges could be used for investigating various problems of blast effect on an electrohydrodynamic analogue computer.

In the second phase of blast effect, particles in the medium would begin to move after attaining initial velocities, and their displacements u_x, u_y, u_z during a certain period of time would be proportional to their velocities. Since the displacements u_x, u_y, u_z would generally not be the same at different points in the medium, the medium would be subjected to deformation. The medium behaves like an elastic one under minor deformations. Under major deformations, irreversible deformations in the form of plastic deformations or deformations of rupture occur, which could lead to medium disintegration. Evidently, disintegration would occur only in the case when the density of deformation energy A of the medium attains a certain limit. For plastic deformations, the value of ultimate specific potential energy may be determined according to the relationship

$$A_s \sigma_s^2/(2E),$$

where σ_s is yield strength; E the modulus of elasticity of the medium.

Kinetic energy equivalent to this, which is more convenient for further computations, is determined by the formula

$$A_s = (E_k)_s = \rho v_s^2/2. \quad \ldots (4.41)$$

By equating these expressions we can obtain the value of critical displacement velocity v_s,

$$v_s = \sigma_s / \sqrt{\rho E}. \quad \ldots (4.42)$$

Brittle failure in the form of fractures from the local rupture of linkages between adjacent particles is characteristic for most rocks. The necessary condition for this is that the stresses have to exceed the ultimate rupture strength of the medium σ_{rup}. But as such a level of stresses would occur in the elastic stage of loading, we have to impart potential energy to the medium, not less than the value $A_{rup} = \sigma_{rup}/(2E)$. This energy can be expressed in terms of critical rate of rupture, similar to (4.41):

$$v_{rup} = \sigma_{rup} / \sqrt{\rho E}.$$

Thus, in that field of the medium where total compression occurs, disintegration would take place only in that part where $A \geq A_s$. Should tensile deformations take place in the rarefied zone, there might be failure under the condition $A \geq A_{rup}$.

Considering the above, we can determine statistically the minimum dimensions of the failure (breakage) zone in which initiation of cracks is highly probable.

O.E. Vlasov and S.A. Smirnov have shown that the total kinetic energy E_K of deformation in a given field can be determined by the expression

$$E_K = \frac{\rho}{2} \iiint \left[\left(\frac{\partial^2 \varphi}{\partial x^2} x + \frac{\partial^2 \varphi}{\partial x \partial y} y + \frac{\partial^2 \varphi}{\partial x \partial z} z \right)^2 + \left(\frac{\partial^2 \varphi}{\partial x \partial y} x + \frac{\partial^2 \varphi}{\partial y^2} y + \frac{\partial^2 \varphi}{\partial y \partial z} z \right)^2 + \left(\frac{\partial^2 \varphi}{\partial x \partial z} x + \frac{\partial^2 \varphi}{\partial y \partial z} y + \frac{\partial^2 \varphi}{\partial z^2} z \right)^2 \right] dx\, dy\, dz. \quad \ldots (4.43)$$

For a field in the form of a cube with face $2a$, after transformation and taking a definite integral along all the axes from $-a$ to $+a$, we can obtain a numerical expression for (4.43) in the following form

$$E_K = \frac{4}{3} a^5 \rho \left[\left(\frac{\partial^2 \varphi}{\partial x^2} \right)^2 + \left(\frac{\partial^2 \varphi}{\partial y^2} \right)^2 + \left(\frac{\partial^2 \varphi}{\partial z^2} \right)^2 + 2 \left(\frac{\partial^2 \varphi}{\partial x \partial y} \right)^2 + 2 \left(\frac{\partial^2 \varphi}{\partial y \partial z} \right)^2 + 2 \left(\frac{\partial^2 \varphi}{\partial z \partial x} \right)^2 \right].$$

The expression within brackets is termed the breakage (failure) criterion and is indicated by Br.

$$Br = \left(\frac{\partial^2 \varphi}{\partial x^2} \right)^2 + \left(\frac{\partial^2 \varphi}{\partial y^2} \right)^2 + \left(\frac{\partial^2 \varphi}{\partial z^2} \right)^2 + 2 \left(\frac{\partial^2 \varphi}{\partial x \partial y} \right)^2 + 2 \left(\frac{\partial^2 \varphi}{\partial y \partial z} \right)^2 + 2 \left(\frac{\partial^2 \varphi}{\partial z \partial x} \right). \quad \ldots (4.44)$$

Then the energy expended for deforming the cube with face $2a$ becomes

$$E_K = 4a^5 \rho Br/3. \quad \ldots (4.45)$$

It is evident from (4.44) that the failure criterion Br is a function of the co-ordinates. Hence deformation energy depends not only on the location of the field under study, but also on the dimension of field a, according to expression (4.45). This relationship is determined by half of the cube's side raised to the power five and not to the power three, reflecting the proportionality E_K to volume of the particular field (cube). This means that for larger fields the energy E_K would significantly exceed the work necessary for attaining plasticity as expressed by equation (4.41),

$$(E_K)_s = 8a^3 A_s = 4a^3 \rho v_s^2.$$

If the field is of smaller dimensions, the value of E_K would be very small and no irreversible processes would take place. Obviously, the more probable dimension of field a, which should remain undisturbed with no further failure, is established from the condition $E_K = (E_K)_s$ or $\frac{4}{3} a^5 \rho \, Br = 4a^3 \, \rho v_s$, hence

$$a = \sqrt{3} v_s / \sqrt{Br}. \qquad \ldots (4.46)$$

For brittle rock failure the value of v_{rup} should be substituted for v_s in (4.46).

It follows from (4.46) that at larger values of Br the values of a are small and vice versa. Consequently, the value of Br would somewhat determine rock failure and the degree of its fragmentation.

Expressions (4.44) and (4.46) enable one to evaluate the more probable field (fragment) which should remain intact, without further disintegration. It suffices if we determine the value of the breakage factor Br at a given distance from the charge, using an analogue computer EHDA, and find values of the critical velocities v_s or v_{rup}. It is very easy to obtain the value of Br for a blast of spherical charge, as its value can be readily computed based on spherical symmetry in movement of the medium induced by the blast. After simple calculations, O.E. Vlasov and S.A. Smirnov obtained this expression for the failure criterion $Br = 3R_{0_z} Q/(\pi \rho r^6)$ and probable size of rock fragment

$$\tilde{a} = v_s r^3 \sqrt{\pi \rho / (R_{0_z} Q)}. \qquad \ldots (4.47)$$

The expression obtained for probable size of rock fragments shows that with an increase in distance from the charge centre, the size of fragment increases, but at one and the same distance the greater the radius of charge and blast energy, the lower the fragment size. This aspect completely accords with the physics of the process and practical experience in failure of rocks by blasting. Obviously, very small fragments are observed very close to the charge. A fragment located at the extreme size $2a$, would indicate that the distance between centres of fragments and charge was

$$r = R_{0_z} + a, \quad a = r - R_{0_z}. \qquad \ldots (4.48)$$

To determine the size of fragments, therefore, equations (4.47) and (4.48) should be solved together. The former is a parabola of the third order in the system of co-ordinates (r, a), while the latter is a straight line inclined at an angle 45° to the axis of r (Fig. 48, a). From the curves shown in the figure it can be seen that in the sector of positive r there would be either two points of intersection or none, while at the extreme where the straight line AB becomes tangential to the parabola, one common point. Obviously, this case is the condition for probable breakage of rock. Thus the equation indicating a failure condition would be

$$\partial a/\partial r = 3v_s r^2\sqrt{\pi\rho/(R_{0_z}Q)} = 1, \qquad \ldots (4.49)$$

which in general can be written as $\partial a/\partial n = 1$, where n is normal to the charge surface.

From formulae (4.47) and (4.48), we get $3r^2 = r^3/a$; $r = 3a$. It is evident from this and from equation (4.48), that $a = R_{0_z}/2$ and substituting the value $r = 3a = \frac{3}{2}R_{0_z}$ in (4.49), we get

$$\frac{9}{4}R_{0_z}^2 = \frac{1}{3v_s}\sqrt{\frac{R_{0_z}Q}{\pi\rho}},$$

whence, $Q = \frac{729}{16}\pi\rho v_s^2 R_{0_z}$. For the given quantum of blast energy, fragmentation is possible with fewer lumps of $2\,a = d_z$ in size. If the blast energy is more, fragmentation occurs with the formation of finers; The minimum lump size would obviously be at $r = R_{0_z}$:

$$a_{min} \geq v_s R_{0_z}^{5/2}\sqrt{\pi\rho/Q}.$$

The expression for the maximum size of fragment and its location relative to the charge centre can be established by drawing a tangent to the cubic parabola (4.47). We get directly from equation (4.49) and (Fig. 48, b)

$$r = (3v_s)^{-1/2}\left(\frac{R_{0_z}Q}{\pi\rho}\right)^{1/4}.$$

Substituting this expression in equation (4.47), we get

$$a_{max} = 3^{-3/2}v_s^{-1/2}(R_{0_z}Q/\pi\rho)^{1/4}.$$

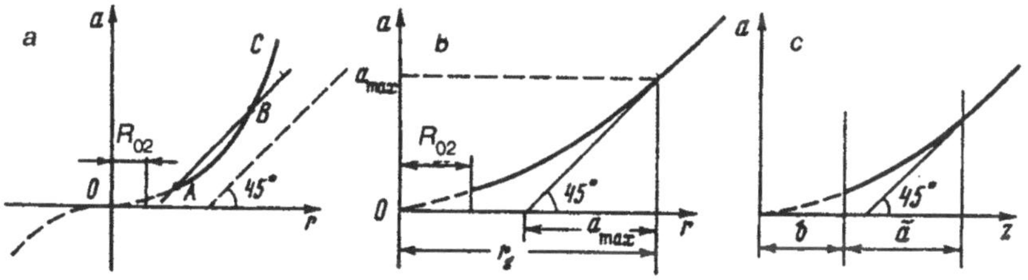

Fig. 48. Curves for (a) computing breakage zone; (b) establishing dimensions of breakage zone of a spherical charge; (c) curve for computing fragmentation along the axis of an extended charge.

It follows that the boundary of failure zone would be located at a distance r_s from the blast centre,

$$r_s = r + a_{max} = R_{cr} = (4/\sqrt{27v_s})4\sqrt{R_{0_z}Q/\pi\rho}.$$

It should be noted that generally two different zones of failure are possible: *first,* failure due to plastic (shear) deformation; *second,* crack formation owing to brittle failure (rupture deformations). The latter is determined by the critical speed of rupture v_{rup}, and all other parameters of this zone (size of fragments, zonal boundary) may be determined by the foregoing relationships in which v_{rup}, the critical speed of rupture, should be substituted for v_s. For the zone of crack formation, we obtain:

$$a_{min} = v_{rup}R^{5/2}\sqrt{\pi\rho/Q};$$

$$a_{max} = (R_{0_z}Q/\pi\rho)^{1/4}/\sqrt{27v_{rup}};$$

$$r_{rup} = R_{frac} = \left(4/\sqrt{27v_{rup}}\right)(R_{0_z}Q/\pi\rho)^{1/4}.$$

It is theoretically very difficult to obtain similar relationships for parameters of breakage of an unconstrained medium by blasting cylindrical charges. O.E. Vlasov and S.A. Smirnov obtained such relationships for two cases: along the axis of charge and in the equatorial plane.

In the *first case* (in the direction of z axis), the expression for failure criteria Br is obtained as

$$Br = 24A^2z^2/(z^2 - c^2)^4.$$

The average size of fragments, according to (4.46), in the direction of z-axis would be

$$\tilde{a} = v_{fr}(z^2 - c^2)/(2Az\sqrt{2}). \qquad \ldots (4.50)$$

The boundary of the failure zone is determined from the condition $\partial\tilde{a}/\partial n = 1$. Vlasov and Smirnov investigated an extreme case and established that actually failure was possible when according to (4.48) (Fig. 48, c) $z = b + \tilde{a}$, so the task boils down to solving the following equation, taking into account equality (4.50):

$$2z^3 - 3bz^2 + 2cz - bc^2 = 0.$$

Equation $z = b + (ab)/c^2$ was also solved by Vlasov and Smirnov and an approximate formula derived for the minimum value $A = 2\sqrt{2}a^2v_{fr}$.

In the *second case* (at the equatorial plane) the nature of breakage in the middle plane xy perpendicular to the charge axis was studied and the expression for Br was as follows:

$$Br = \frac{8A^2(3r^4 + 3c^2r^2 + c^4)}{r^4(r^2 + c^2)^3}$$

and the average fragmentation

$$\tilde{a} = \frac{v_{fr} r^2}{4A}\left[\frac{6(r^2+c^2)^3}{3r^4+3c^2r^2+c^4}\right]^{\frac{1}{2}}. \qquad \ldots (4.51)$$

At very high values of r, considering only the leading terms of r under the square root $\sqrt{6r^6/(3r^4)}$, this relationship simplifies to:

$$\tilde{a} = v_{fr} r^3/(2\sqrt{2}A).$$

A similar result is obtained from equation (4.49) at large values of z

$$\tilde{a} = \frac{v_{fr} z^3}{2\sqrt{2}A}.$$

This conclusion accords with the physical representation of an extended charge as a concentrated charge at very large distances.

Boundaries of failure zone are established by equation (4.51) according to the general condition $\partial\tilde{a}/\partial n = 1$, which is obtained as follows:

$$\frac{x(9x^6+12x^4+8x^2+2)\sqrt{x+1}}{(3x^4+3x^2+1)^{3/2}} = B,$$

where $x = r/c$; $B = 2\sqrt{2}A/(\sqrt{3}c^2 v_{fr})$.

O.E. Vlasov and S.A. Smirnov solved this equation for $1 \geq x \geq 0$ (Fig. 49, a).

Taking into account the fact that the choice of rational value of LLR is also related to some extent to the determination of probable cross-sectional dimensions of the breakage zone, we use formula (4.51) for evaluating the minimum dimensions of rock fragments in the equatorial plane, by putting $r = a$ and considering that $b \gg a$. As a result, we get

$$a_{min} = \frac{d_z^2 v_{fr}}{4}\left[\frac{3\pi\rho}{2q}\ln\left(\frac{l_z}{d_z}\right)\right]^{1/2}.$$

This relationship confirms the very decisive influence of charge diameter on the fine crushing of rocks in a blast, since with an increase in d_z, a_{min} also increases, which means that the volume of finer rock fragments decreases. Formula (4.51) enables one to roughly assess the granulometric composition of the broken part of rock. Its precise calculation can be done by using the method of EHDA for investigating the distribution of rocks according to their size over the entire meridinal section.

Assuming that the surfaces of equal sized rocks around an extended charge approximate to cylindrical surfaces, O.E. Vlasov and S.A. Smirnov established the granulometric composition only along the middle section. Assuming that volume V of all the fragments whose size does not exceed a is approximately $V = \pi l_z(r^2 - a^2)$, and hence $r^2 = a^2 + V/(\pi l_z)$, we get an expression for (4.50) in the following form,

$$G = \xi \left[\frac{(\xi + 1)^2}{3\xi^2 + 3\xi + 1} \right]^{1/2}, \qquad \ldots (4.52)$$

where $\xi = V/(\pi l_z b^2)$; $G = 4A\tilde{a}/(6b^3 v_{fr})$.

Expression (4.52) thus obtained would represent an approximate curve of probable granulometric composition, which for ξ varying between $1 \geq \xi \geq 0$ is shown in Fig. 49, b.

While studying this problem using the method of EHDA, it should be borne in mind that electric potential is distributed as per its scaled value but not as per the absolute, i.e., potential as a ratio of the total drop in voltage on the model. The value of scaled potential on the model surface is taken as equal to unity but on the free face as equal to zero. Here the absolute values of potential at various points are determined by multiplying their relative (scaled) values with the absolute value of velocity potential on the charge surface, which is generally computed by the energy balance, and for an extended charge $b \gg a(2b = l_z, 2a = d_z)$, is described by formula (4.40) in the case of unconfined medium. The total quantum of kinetic energy imparted to a medium from a blast is described by formula (4.35).

Current passing through the model in the EHDA device is determined by the relationship

$$I = \frac{1}{\tau} \oint \frac{\partial U_{surf}}{\partial n}\, dt = \frac{U_{surf} - U_0}{R},$$

where τ is the specific electrical resistance of the medium used in the model; R the electrical resistance of the model; U_{surf}, U_0 voltage (potential) at the charge surface and free face, respectively.

Putting $U_0 = 0$ and $U_{surf} \approx \varphi_{surf}$ from (4.52) and (4.37), we get

$$\frac{U_{surf}\tau}{R} = \frac{2Q}{\rho\varphi_{surf}}; \quad \varphi_{surf} = U_{surf}\sqrt{\frac{2Q}{\rho\tau}}\sqrt{R}. \qquad \ldots (4.53)$$

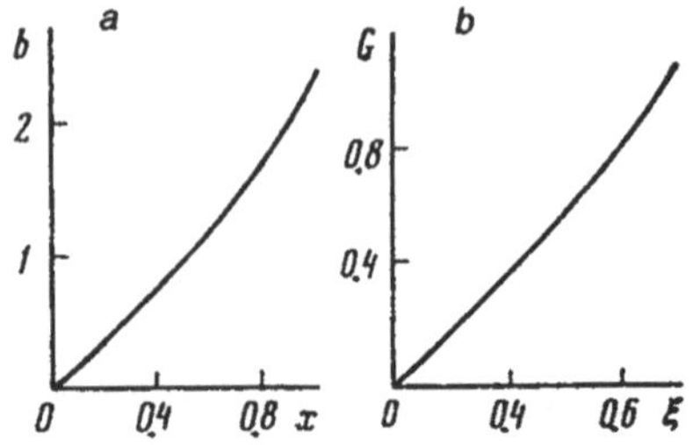

Fig. 49. Curves for determining (a) boundary of breakage zone and (b) granulometric composition in a blast of extended charge.

Comparing (4.40) and (4.53), it can be seen that (potential) velocities on the surface of charges possessing the same energy, but with different ratio l_z/d_z, would be related in terms of $\sqrt{R}$ measured for models of these charges.

In simulating failure by the EHDA method, apart from plotting equipotential curves as per the methodology described in Sec. 4.1.1 and determining the velocities of displacement of particles in the medium by formula (4.34), it is necessary to compute along co-ordinates x, y, z the second derivatives of velocity potential φ contained in the expression for failure criteria Br. This can be done by a square grid (Fig. 50). These derivatives are computed by the formula:

$$\left(\frac{\partial\varphi}{\partial x}\right)_1 = \frac{\varphi_{22} - \varphi_{21}}{a}; \quad \left(\frac{\partial\varphi}{\partial x}\right)_2 = \frac{\varphi_{23} - \varphi_{22}}{a};$$

$$\left(\frac{\partial^2\varphi}{\partial x^2}\right)_0 = \frac{\left(\frac{\partial\varphi}{\partial x}\right)_2 - \left(\frac{\partial\varphi}{\partial x}\right)_1}{a} = \frac{\varphi_{23} - 2\varphi_{22} + \varphi_{21}}{a^2};$$

$$\left(\frac{\partial^2\varphi}{\partial y^2}\right)_0 = \frac{\varphi_{12} - 2\varphi_{22} - \varphi_{32}}{a^2};$$

$$\left(\frac{\partial^2\varphi}{\partial x\,\partial y}\right)_0 = \frac{\varphi_{12} - \varphi_{11} - \varphi_{23} - \varphi_{31}}{4a^2}.$$

In doing so it is necessary to record precisely the distribution of velocity potential at a large number of points, which significantly complicates the computational work. In order to reduce the difficulty in computation, the scale of modelling should be increased, which also results in reduction of

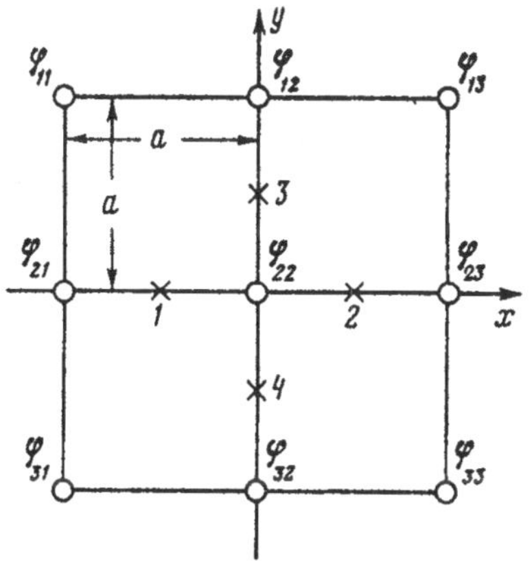

Fig. 50. Diagram depicting methodology for computing breakage criteria in order to evaluate fragmentation when an extended charge is blasted in any direction relative to its axis.

relative errors in recording the distribution of potentials. Further, to reduce the difficulty in determining the degree of fragmentation while simulating by the EHDA method, one should attempt to measure directly the second derivatives of the velocity potential. It is also necessary during simulation to maintain the geometric similitude of all linear dimensions in the model and the full-scale. The following phenomena are used for simplifying simulation and for reducing laborious operations in the EHDA method.

If a flat surface (wire, line) of a dielectric or a surface in contact with air is placed near an electrode, simulating the charge, the electric field will be similar to that generated in the presence of a second electrode (charge) in the medium being simulated, charged to the same extent as that of the first but placed away from it at a distance equal to twice the distance from the electrode to this surface (wire, line). If in place of the dielectric we place a flat surface (wire, line) of a conductor, a field would also be formed due to the two electrodes situated at the same distance from the surface, but having two opposing charges, i.e., mirror image from the free face is simulated.

These properties of dielectric and conducting surface (wire, line) enable one to study not the entire circuit, but only that part of it confined between planes of symmetry when electrodes (model) are symmetrically placed. As an example, the distribution of velocity potential for a model of an extended (shot hole) charge perpendicular to the flat free face is shown in Fig. 51. A brass rod (strip in a flat model), corresponding to the geometric scale of dimensions, placed perpendicular to a flat copper sheet (widthwise in a flat model) at a distance equal to the stemming length, is chosen as the model in these figures. In this case stemming has the same properties as that of the basic rock. Equipotential curves are plotted according to the methodology described in Sec. 4.11, based on which velocity can be computed at any point in the field. While doing so the direction of velocity should coincide with the normal to the equipotential curve at this point. The density of the equipotential curves would characterise the density of kinetic energy around the charge in the medium while direction of vectors of velocity indicates direction of movement of the medium at the given points.

The initial velocity field obtained in a blast of single shothole charge is shown in Fig. 51, a. Here point *M* indicates the possible boundary of a blast cone. The initial velocity field of the model simulating the cut obtained by firing a dummy and charged (shaded) shot holes is shown in Fig. 51, b. In this case the maximum density of equipotentials (maximum density of energy) occurs between the shot holes whose velocity vectors are directed towards the empty hole; this indicates favourable conditions for breakage of the barrier between the shot holes. Here the cone will have a skewed shape if the blast energy is adequately large, i.e., a huge portion of the rock will be ejected from the field of the charged shot hole. Finally, the initial velocity field of the model simulating the cut obtained by simultaneously blasting two

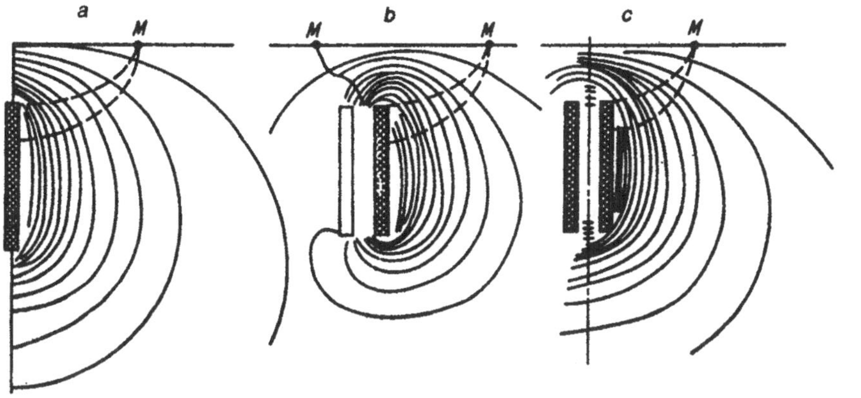

Fig. 51. Initial velocity field of model under the effect of (a) single shothole charge; (b) dummy and charged (shaded) shot holes; (c) two charged shot holes.

charged shot holes is shown in Fig. 51, c. In this case the energy density between shot holes is low and particles in the medium do not move (an intense compression takes place). The possible cone in this case would be larger by the magnitude of the distance between charges, when its depth remains constant. This last conclusion regarding the height of the cone does not take into account the fact of increase in the effective radius of converging charges, in conformity with which there would also be an increase in depth of cone due to an increase in effective charge radius.

4.3.2 Possibility of Applying the Method of Hydrodynamic Analogy for Simulating Various Processes of Rock Fragmentation by Blasting and its Future Prospects

Let us first highlight the fact that the method of hydrodynamic analogy (EHDA) can describe only the wave effect of a blast. Assuming that as a result of blasting an ideal elastic medium receives certain kinetic energy from particle movement in the medium, a potential field of velocities with potential φ is created in the medium. Here the charge is characterised only by thickness and shape while it is instantly primed, i.e., the process of blasting amounts to an instantaneous release of explosive energy over the entire volume of the charge. Such a formulation of the problem does not at all take into account time-related characteristics of the detonation process of the charge, progression of blast and propagation of waves in the medium and its fragmentation, nor the piston effect of detonation products which, evidently, plays a key role in questions related to displacement of rock mass, formation of blast cone and muck pile.

The method of simulating rock fragmentation by a blast discussed above does not consider either the breakage mechanism nor the physicomechanical properties of the massif. It is based solely on the critical velocities of rock displacement under compression v_s and under tension v_{rup} that characterise respectively plastic and brittle breakage of these rocks and is determined by the yield limit σ_s and ultimate tensile strength σ_{rup} of rocks as computed by formula (4.42) and (4.43). Moreover, it is assumed that the entire energy of blast is fully converted to kinetic energy of rock movement [see formula (4.35)], neglecting the dissipative losses of blast energy in the near-field zone of the blast as well as the energy of detonation products remaining within the blast cavity. All these assumptions significantly reduce the reliability of results obtained by this method and do not permit direct extrapolation of the results of simulation to the full-scale. Nevertheless they evoke a certain interest for comparative qualitative assessment of the stress field when certain conditions (parameters) of blasting are altered. In particular, one can study the effect of spacing between charges fired simultaneously on the field of velocities in rock considering the free face to be parallel to the row of charges. In this case a flat model can be used, which conforms to the case of using infinitely long charges parallel to the free face. To simplify the experiment, the symmetry conditions of the circuit indicated in Sec. 4.3.1 can be adopted by making use of strips made from dielectric and conducting materials. The strip (line, wire) made from the conductor is installed on the free face line, while that made from the dielectric material is placed in the middle of charges perpendicular to the free face. Fields of velocity potential (Fig. 52) for varying distances between charges, corresponding to the proximity coefficients $m = 0.4$; 0.8; 1.2; 1.6 and 2.4, were obtained on the EHDA installation at a scale of $l_c = 25$ when explosive charges with a thermal energy of explosion $Q = 4190$ kJ/kg were fired in granite having density $\rho_p = 2600$ kg/m^3 and critical velocity of break-up $v_{fr} = 1.4$ m/s. The blast holes were 200 mm in diameter and charge length 10 m at LLR of 10. Here the energy of charge was computed by

$$Q_z = \frac{\pi d_z^2}{4} l_z \gamma_{EC} \, Q = 13.16 \times 10^8 \text{ kJ},$$

and velocity potential at the charge surface $\varphi_{surf} = 272$ m^2/s.

In Fig. 52, a–e, the relative values of potential of the equipotential curves vary from 0.95 (first charge from the curve) to 0.05 at intervals of 0.05. Absolute values of φ are obtained by multiplying them with $\varphi_{surf} = 272$ m^2/s. The curves of distribution of fragments along the LLR and along the line of charges respectively are shown in Fig. 53. It follows from these curves that the best fragmentation is obtained in the direction of LLR at $m = 1.6$ and the worst at $m = 0.8$; in the direction along the line of charges the best fragmentation occurs at $m = 1.2$ and 1.6 and the worst at $m = 2.4$. At the

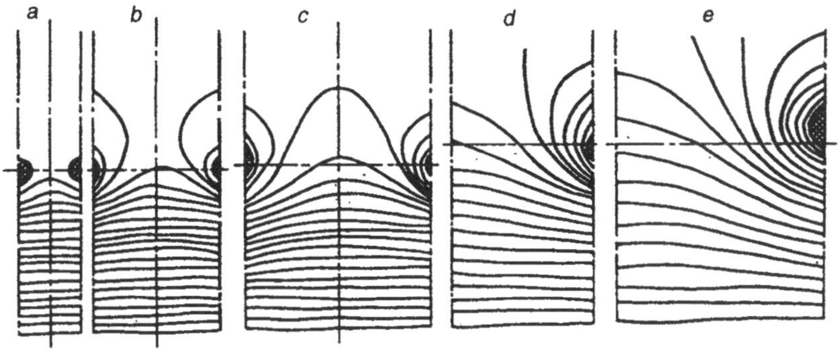

Fig. 52. Fields of velocity potentials recorded from the EHDA simulator when a row of charges of different proximity coefficients *m* is fired.

a—$m = 0.4$; b—$m = 0.8$; c—$m = 1.2$; d—$m = 1.6$; e—$m = 2.4$.

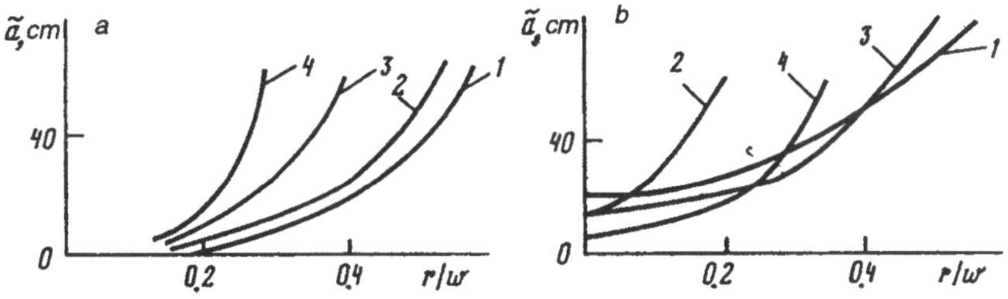

Fig. 53. Curves showing distribution of fragments for various proximity coefficients. a—along the line of LLR; b—along the centre line passing between charges; 1, 2, 3, 4—respectively at $m = 1.6$, 2.4, 1.2 and 0.8.

least proximity coefficient $m = 0.4$, equipotential lines away from the charges represent parallel straight lines placed at the same distance between each other, which confirms the possibility of failure only nearer to the charges and the translational motion of medium in the direction of LLR but at a distance away from it. The results obtained agree, to some extent, with the data of other researchers and field blasts. In particular, for granites the optimal coefficient of proximity of charges [3], based on the action of stress wave, varies within the limits $m = 2$ and $m = 3$ depending on the Poisson ratio of rock $\nu = 2$ or 3.

The nature of equipotentials at the middle portion between the charges shows that direction of velocities of medium movement at this portion facilitates initiation and formation of a cavity between charges. As per the magnitude of velocity component v_y from these curves, one can judge the possibility of cavity formation at a given spacing between charges, after verifying the condition $(v_y)_{min} \geq v_{rup}$. Here the authenticity of results would depend on correctness of evaluating the blast energy spent towards the kinetic energy for moving the medium, and magnitude of critical velocity of rupture, which can be suitably computed by the relationship

$$v_{rup} = [\sigma_{dn.rup}]/(\rho_p c_p),$$

where $[\sigma_{dn.rup}]$ is the dynamic tensile strength of rock; $\rho_p c_p$ the acoustic stiffness of rock.

Simulation by the EHDA method can also be adopted for investigating the blast effect in stratified media, which is significant in practice for computing parameters of drilling-blasting operations (DBO) in the working of narrow-vein ore bodies, when the zone of breakage of enclosing rocks should be minimal. In such a case the stratified medium is simulated by model media having different degrees of electric conductivity in corresponding analogous regions. However, fabrication of such models is wrought with certain difficulties. O.E. Vlasov and S.A. Smirnov suggested a simpler method for fabricating the model of a stratified medium from a medium of homogeneous conductivity, in which for creating a sharp rise in potential at the boundary between the layers it is necessary to introduce a resistance satisfying the condition

$$\gamma_1 \varphi_1 = \gamma_2 \varphi_2, \qquad \ldots (4.54)$$

where γ_1 and γ_2 are volumetric weights at the contour of layers; φ_1 and φ_2 values of velocity potential at two adjacent layers for the case $\varphi_2 < \varphi_1$ and $\gamma_2 < \gamma_1$.

In this case the potential φ within each layer satisfies Laplace's equation $\nabla\varphi = 0$, while the normal derivatives to the boundary separating the layers, due to the incompressibility of media, are taken to be equal, i.e., $\partial\varphi_1/\partial n = \partial\varphi_2/\partial n$ (n—direction of normal to the boundary separating the media). When the specific resistance at the surface of separation is R, the current flux passing between these layers is given by,

$$(\varphi_1 - \varphi_2)/R = \frac{1}{\tau}\frac{\partial\varphi}{\partial n}, \qquad \ldots (4.55)$$

where τ is the specific resistance of the medium used in the model.

An expression can be obtained for resistance from formulae (4.54) and (4.55)

$$R = -\tau\frac{\varphi_1 - \varphi_2}{\partial\varphi/\partial n} = \left(\frac{\gamma_2}{\gamma_1} - 1\right)\frac{\tau\varphi_2}{-\partial\varphi/\partial n}.$$

This resistance can be made, for example, from silk or cotton thread impregnated with current-conducting paint made from graphitic powder mixed with gum of the type BF, whose electric conductance can be altered within wide limits, by varying the amount of graphite added to the paint.

This method was also adopted for modelling fissured rocks, by simulating the system of cracks by a geometrically similar grid of such threads. The foregoing is applicable to the resistance of a thread simulating a crack R_T, i.e., the following relationship can be utilised:

$$\gamma_1\varphi_1 = \gamma_T\varphi_T;\ \partial\varphi_1/\partial n = \partial\varphi_T/\partial n = -\frac{1}{\tau}\frac{\partial\varphi_1}{\partial n}, \qquad \ldots (4.56)$$

where φ_T and γ_T are potential and density of the crack filler respectively; R_T resistance simulating the crack in the model.

Assuming for the sake of simplicity the presence of a single crack of width δ, located at a distance r from the axis (centre) of the charge, we have in analogy with the foregoing discussion, for the region up to the crack $\varphi_1 = A + \frac{k}{r}$ for spherical or axial symmetry in the equatorial plane. Taking $\varphi_1 = 1$ at the surface of the charge, we have $A = 1 - k/R_{0_z}$, where R_{0_z} is the charge radius, while the potential at distance r_1 would be $\varphi_1 = 1 + k(1/r - 1/R_{0_z})$. Hence the normal derivative $\partial\varphi_1/\partial n = -k/r^2$. At the boundary between crack and the post-crack region, the velocity potential $\varphi_2 = B + k/(r+\delta)$. Considering that even at the boundary between the crack and the second region the equality $\gamma_T\varphi_T = \gamma_1\varphi_2$ holds good, we can express φ_T in terms of φ_2 according to the relationship,

$$\varphi_T = \varphi_2\frac{\gamma_1}{\gamma_2} = \frac{\gamma_1}{\gamma_2}\left(B + \frac{k}{n+\delta}\right).$$

The constant B can be determined from the condition φ_2 at the free face in a manner similar to that of A for φ_1 from the condition at the charge surface. Obviously at $r = W$, $\varphi_2 = 0$ and $B = -k/W$.

Then,

$$\varphi_T = \frac{\gamma_1}{\gamma_2}\left(\frac{k}{r+\delta} - \frac{k}{w}\right) = \frac{\gamma_1}{\gamma_2}k\left(\frac{1}{r+\delta} - \frac{1}{w}\right). \qquad \ldots (4.57)$$

From the first condition of formula (4.56) $\varphi_1 = \varphi_T\gamma_T/\gamma_1$, and the last condition of this formula, we have

$$R_T = \frac{(\varphi_1 - \varphi_2)\tau}{-\partial\varphi/\partial n} = \frac{r\tau}{k}(\varphi_1 - \varphi_T) = \frac{r^2\tau}{k}\varphi_T\left(\frac{\gamma_T}{\gamma_1} - 1\right). \qquad \ldots (4.58)$$

Substituting the value of φ_T from (4.57) in equation (4.58) we have

$$R_T = \frac{r^2\tau}{k}\left(\frac{\gamma_T}{\gamma_1} - 1\right)\frac{\gamma_1}{\gamma_T}k\left(\frac{1}{r+\delta} - \frac{1}{W}\right)$$
$$= \tau r^2\left(1 - \frac{\gamma_1}{\gamma_T}\right)\left(\frac{1}{r+\delta}\right)\left(\frac{1}{r+\delta} - \frac{1}{W}\right).$$

It has been shown in [8] that the effect of a crack on stress wave amounts to a reduction in amplitude of the stress wave, barely altering its pulse, and is determined by the total width of all cracks through which the wave travels. Consequently, jointing can be approximately assessed by a single crack having a width equal to the total width and situated roughly at the middle between the charge axis (centre) and the LLR, i.e.,

$$r = W/2; \ \delta = \sum_{i=1}^{n} \delta_i,$$

and γ_T density of material filling the crack. When different fillers are used, the average γ_T is computed as

$$\gamma_T = \frac{\sum_{i=1}^{n} \gamma_{T_i} \delta_i}{\sum_{i=1}^{n} \delta_i}.$$

4.4 Examples for Simulation of Wave Motion by Using Electric Networks and Computers

4.4.1 Solving Ordinary Differential Equations on an Analogue Computer (ABM)

Let us consider the process of solving general differential equations, by the structured network of ABM, taking for example the problem posed by Cauchy:

$$y^{(n)} - F\left[y^{(n-1)},\ y^{(n-2)}, \ldots, y^{1},\ y, t\right] = 0,$$

$$y_{(0)}^{n-1} = y_0^{(n-1)}, \ldots, y'(0) = y'_0,\ y(0) = y_0. \qquad \ldots (4.59)$$

This initial equation is resolved relative to the highest order derivative, i.e., is brought to the form

$$y^{(n)} = F\left[y^{(n-1)},\ y^{(n-2)}, \ldots, y',\ y, t\right] \qquad \ldots (4.60)$$

and for this part of the structural network is plotted in the form of a chain of sequentially connected integrators (Fig. 54, a), at whose entry the corresponding derivatives $y^{(n)}$, $y^{(n-1)}$ etc. up to y' are fed.

A network is developed for realising the right side of equation (4.60), whose concrete form is established by the analytical expression $F(y^{i}t)$. It is indicated by a rectangle with corresponding legend in the network (Fig. 54, b). The outlet of this network is connected to the entry of the first integrator in the chain, which is reflected by equation (4.60).

The corresponding initial conditions are assigned to all the integrators in the form of arrows with indication of initial conditions (Fig. 54, c), thereby completing the solution to the initial differential equation.

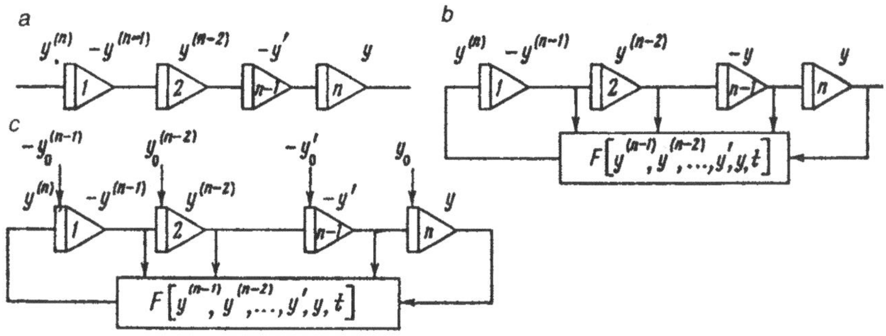

Fig. 54. Structural network of ABM for solving ordinary differential equations of *n*th order. a—chain of integrators; b—introducing right side of eqn (4.60) into this chain; c—introducing initial conditions into respective integrators.

As an example, let us consider the solution to a linear non-homogeneous differential equation of the second order having constant coefficients

$$y''(t) + a_1 y'(t) + a_0 y(t) = f(t),$$
$$y'(0) = y_0', y(0) = y_0. \qquad \ldots (4.61)$$

Rewriting equation (4.61) in the form

$$y''(t) = -a_1 y'(t) - a_0 y(t) + f(t), \qquad \ldots (4.62)$$

we represent it in the form of a chain of integrators (Fig. 55, a).

Let us develop a network for the right side of transformed equation (4.62). Obviously, the network should consist of an invertor (1), summator (2) and block generating the function $f(t)$, such that at the exit of the summator, the sum is given out (right side):

$$-a_1 y'(t) - a_0 y(t) + f(t) = F[-a_1 y' - a_0 y + f(t)]$$

(Fig. 55, b), which then goes to the first integrator, closing the equation being solved.

Finally, the initial conditions are applied to the corresponding integrators y_0' and y_0 (Fig. 55, c). This network enables one to study the variables $y''(t)$, $y'(t)$ and $y(t)$. If we are not interested in the behaviour of $y''(t)$, the network is significantly simplified (Fig. 55, d) by combining the summator (2) and integrator (1) into an integrator-summator (1′).

When variable coefficients $a_1(t)$ and $a_0(t)$ are present in equation (4.62), the scheme of the right side becomes complicated due to the induction of blocks of variable coefficients and blocks of multiplication (Fig. 55, e).

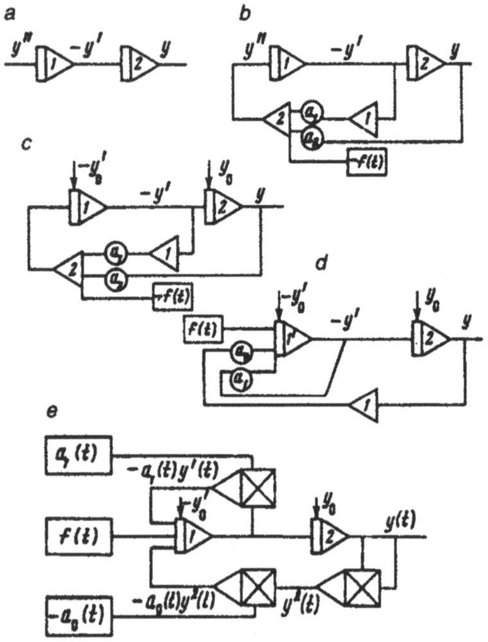

Fig. 55. Structural network for generating on the ABM the solution to a linear non-homogeneous differential equation of second order having constant coefficients.
a—chain of integrators; b—assigning right side of equation; c—inputing initial conditions—closing the circuit; d—transformed network, excluding analysis of behaviour of second derivative function; e—structural network for differential equation of second order with varying coefficients.

4.4.2 Generating Solutions to Differential Equations in Partial Derivative Form Using an ABM

A solution to differential equations in partial derivative form on an ABM can be generated approximately by realising solutions to the system of ordinary differential equations, which approximate the initial equations to partial derivatives. Let us illustrate this through a very simple boundary problem, the Dirichlet problem. For approximation here the partial derivatives are replaced by certain expressions containing only specific values of functions, i.e., formulae pertaining to numerical differentiation. These formulae are very easily found by the method of rectangular grids at equidistant values of the argument.

Let us adopt this method for a certain function $u(x)$ of a variable x. Let for discrete values of x at an interval of h

$$x_{-1} = x_0 - h;\ x_0 \text{ and } x_{+1} = x_0 + h$$

the values of function $u(x)$ are given as: $u(x_{-1})$, $u(x_0)$, $u(x_{+1})$. Expanding the functions $u(x_{-1})$ and $u(x_{+1})$ into a Taylor series, and later subtracting and adding them [14], we can obtain corresponding values for the functi ns $u(x_0)$ in the form

$$u^{I}(x_0) = \frac{1}{2h}[u(x_{-1}) - u(x_{+1})] - \frac{h^3}{6}u^{III}(\varepsilon)\ldots,$$
$$u^{II}(x_0) = \frac{1}{h^2}[u(x_{-1}) - 2u(x_0) + u(x_{+1})] - \frac{h^2}{12}u^{IV}(\varepsilon), \quad \ldots (4.63)$$

where $x - h < \varepsilon < x + h$. Both these expressions yield values of the derivative with a precision of the order h^2 and in case the function $u(x)$ is in the form of a polynomial of the second order, exact values of the derivative are obtained, as $u^{III}(x)$ and $u^{IV}(x)$ in this case are equal to zero.

Similarly, we can determine the derivative functions for other points, x_{-1} and x_{+1}. It has been proved [14] that for equidistant values of the arguments

$$x_0,\ x_0 + h = x_1,\ x_2 = x_0 + 2h, \ldots, x_i = x_0 + ih.$$

The corresponding values of functions and their derivatives

$$u_0 = u(x_0);\ u_1 = u(x_1);\ u_2 = u(x_2), \ldots, u_i = u(x_i);$$
$$u_0^{I} = u^{I}(x_0);\ u_1^{I} = u^{I}(x_1);\ u_2^{I} = u^{I}(x_2), \ldots, u_i = u^{I}(x_i);$$

are related for three values (unidimensional case)

$$u_0^{I} = \frac{1}{2h}[-3u_0 + 4u_1 - u_2] + \frac{h^2}{3}u^{III}(\varepsilon_0);$$
$$u_1^{I} = \frac{1}{2h}[-u_0 + u_2] - \frac{h}{6}u^{III}(\varepsilon_1);$$
$$u_2^{I} = \frac{1}{2h}[u_0 - 4u_1 + 3u_2] + \frac{h}{3}u^{III}(\varepsilon_2),$$

where $x_0 < \varepsilon_0 < x_2$; $x_0 < \varepsilon_1 < x_2$; $x_0 < \varepsilon_2 < x_2$.

We can obtain derivatives in a similar manner even for a large number of points. It has been shown [14] that all the formulae of differentiation can be represented approximately in the form

$$\vec{u}^{I} = \frac{1}{h}M\vec{u};\ \vec{u}^{II} = \frac{1}{h}M\vec{u}^{I} = \frac{1}{h^2}M^2\vec{u}, \qquad (4.64)$$

where $\vec{u}^{II}, \vec{u}^{I}$ and $\vec{u}$ are n-dimensional vectors, M the square matrix, which in the case of three points would be respectively equal to

$$M_3 = \begin{vmatrix} -1.5 & 2.0 & -0.5 \\ -0.5 & 0.0 & 0.5 \\ 0.5 & -2.0 & 1.5 \end{vmatrix}.$$

Computing the values of partial derivatives of several variables differs in no way from computing the values of derivatives of a function of a single variable. Thus, for a function of two variables we can write according to formula (4.64)

$$\vec{u}^{\mathrm{I}}_x = \frac{1}{h_x} M \vec{u}(x,y); \; \vec{u}^{\mathrm{II}}_x = \frac{1}{h_x^2} M^2 \vec{u}(x,y);$$
$$\vec{u}^{\mathrm{I}}_y = \frac{1}{h_y} M \vec{u}(x,y); \; \vec{u}^{\mathrm{II}}_y = \frac{1}{h_y^2} M^2 \vec{u}(x,y);$$
$$[\vec{u}^{\mathrm{II}}(x,y)] = \frac{1}{h_x h_y} M[\vec{u}(x,y)]^T M^T,$$

where $[\vec{u}^{\mathrm{II}}(x,y)]$ and $[\vec{u}(x,y)]$—matrix expressions; and the symbol T denotes the transpose operation.

As an example, let us consider approximation of the equation for thermal conductivity

$$\partial\theta/\partial t = \frac{\lambda \partial^2 \theta}{c \partial y^2}, \qquad \ldots (4.65)$$

depicting variation in temperature θ in a wall of thickness l, dividing two media having temperatures $\varphi_1(t)$ and $\varphi_2(t)$ changing over time (Fig. 56, a). In equation (4.65), λ is the coefficient of thermal conductivity and c the specific volumetric heat capacity of the wall. In order to determine a system of ordinary differential equations that approximate equation (4.65), we shall split the y-axis into a series of points separated by equal distances of h, at each of which the temperature would be a function of time alone. A partial derivative of such a point might be roughly represented through values $\theta(y, t)$ at two points adjacent to it i_{-1} and i_{+1}, i.e., by formula (4.63)

$$\frac{\partial^2 \theta(y_i, t)}{\partial y^2} = \frac{1}{h^2}[\theta_{i+1}(t) + 2\theta_i(t) + \theta_{i-1}(t)].$$

Taking this expression into account equation (4.65) can be written in the form of ordinary differential equations of the first order as

$$\frac{\partial \theta_i(t)}{\partial t} = \mu[\theta_{i+1} + \theta_{i-1} - 2\theta_i];$$
$$\theta_i(0) = T(y_i) = T_i; \; \mu = \lambda/ch^2;$$
$$i = 1, 2, \ldots, n, \qquad \ldots (4.66)$$

where $T(y)$ is a given function featuring the distribution of temperature over wall thickness at the initial period of time; n is the number of points on the y-axis.

The structural network of ABM, generating a solution to the system of differential equations (4.66), is shown in Fig. 56, b where for generating each equation of the system only a single integrator is required, while to generate the entire system a chain of n integrators is required. The networks of ABM generating the given functions of time $\varphi_1(t)$ and $\varphi_2(t)$ are shown as rectangles in the network at the first and last integrators. These functions characterise the temperature in the media surrounding the wall.

While scaling the variables it should be remembered that temperature at the internal points of the wall ought not to exceed values of temperature $\varphi_1(t)$ and $\varphi_2(t)$.

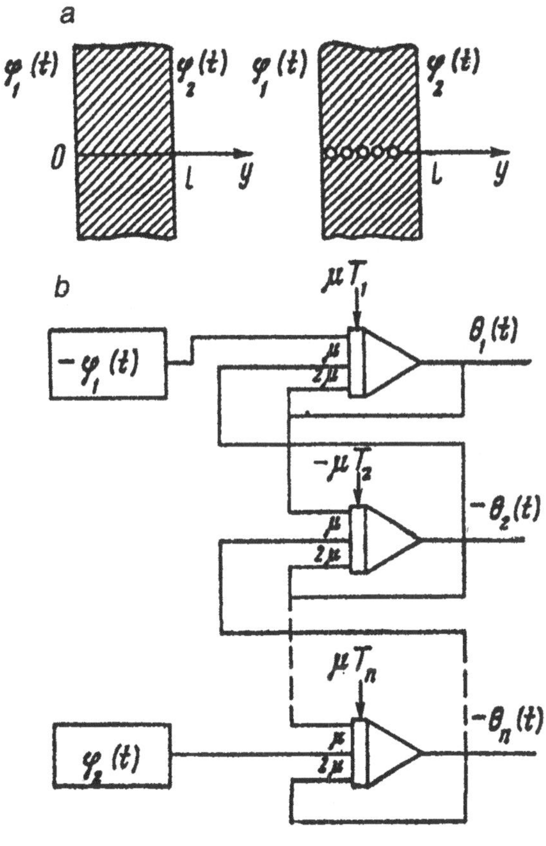

Fig. 56. Structural networks of ABM, generating solutions to differential equation of thermal conductivity.

a—diagram for formulating the problem; b—structural network of ABM consisting of a sequential system of equations as per the number of points obtained on splitting the y-axis.

4.4.3 Mathematical Simulation of Wave Motion Induced by a Blast

Firstly, let us approximate by an ABM the solution of wave equation valid for elastic waves and of the form

$$\frac{\partial^2 u}{\partial x^2} + \frac{\partial^2 u}{\partial y^2} + \frac{\partial^2 u}{\partial z^2} = \frac{1}{c^2}\frac{\partial^2 u}{\partial t^2}. \qquad \ldots (4.67)$$

where u is the displacement of points belonging to the elastic medium, in the wave induced by blasting; x, y, z are co-ordinates of the point; t current time; c velocity of propagation of elastic wave, $c^2 = (\lambda + 2\mu)/\rho$; λ Lamé's constant; μ elastic shear modulus of rock; ρ rock density.

Equation (4.67) may be written for the unidimensional case, in the following form of ordinary differential equation with respect to time t for any point i, by adopting the method of finite differences

$$\frac{\partial^2 u}{\partial t^2} = \frac{c^2}{h^2}[u_{i-1}(t) - 2u_i(t) + u_{i+1}(t)], \qquad \ldots (4.68)$$

where $h = \Delta x$ is the distance between points $i = 1, 2, \ldots, n$ along the line of progression of wave; u_{i-1}, u_i, u_{i+1} are displacements in the wave at adjacent points.

To solve equation (4.68), it is necessary to assign initial and boundary conditions. For example, in a blast at the boundary of an elastic zone at time $t = 0$ [time of reckoning wave progression from this zone $u = u_0(x = 0, t = 0)$], at distances $x \to \infty$, $u_n \to 0$. In this case the structural network of the ABM generating a solution to equation (4.68) can be constructed in accordance with the circuit of approximation described in Secs. 4.4.1 and 4.4.2 in the form of a chain consisting of two integrators (Fig. 57, a), where $\lambda = \alpha^2/h^2$. It can be seen from this figure that values of $u_0(t)\alpha$. $u_2(t)\alpha$ are fed (from the output of integrator 2 of the second cell) to the integrator-summator 1 of the first cell, while the value at the first point $-2\alpha u_1(t)$ is fed to it (1) from the output of integrator 2 of the first cell through inverter 3, i.e. the integrator-summator 1 obtains the signal

$$\alpha[u_0(t) + u_2(t) - 2u_1(t)] = \partial^2 u/\partial t^2.$$

and this means that in the first cell the equation (4.68) for the first point is solved. Similarly, the second cell in the chain solves this equation for the second point etc., up to $(k - 1)$-th point, thereby obtaining at the outlet of A_i the corresponding values of displacement of particles in the elastic zone $u_i(t)$ given out by integrator 2.

As another example, let us consider wave motion in an elastoviscous medium. For unidimensional movement in the direction of the axis, the wave motion is described by an equation of the type

$$(\lambda + 2\mu)\frac{\partial^2 u}{\partial x^2} + \frac{4}{3}\eta\frac{\partial}{\partial t}\left(\frac{\partial^2 u}{\partial x^2}\right) = \rho\frac{\partial^2 u}{\partial t^2}, \qquad \ldots (4.69)$$

where η is the viscosity coefficient of the medium. Differentiating equation (4.69) with respect to time t and denoting $v = \partial u/\partial t$ displacement velocity of particles in the medium, we get

$$(\lambda + 2\mu)\frac{\partial^2 v}{\partial x^2} + \frac{4}{3}\eta\frac{\partial}{\partial t}\left(\frac{\partial^2 v}{\partial x^2}\right) = \rho\frac{\partial^2 v}{\partial t^2}. \qquad \ldots (4.70)$$

For approximating the solution to this equation on the ABM, let us resolve it relative to the higher order derivative in the form

$$\frac{\partial^2 v}{\partial t^2} = \frac{4}{3\rho}\eta\frac{\partial}{\partial t}\left(\frac{\partial^2 v}{\partial x^2}\right) + \frac{\lambda + 2\mu}{\rho}\left(\frac{\partial^2 v}{\partial x^2}\right)$$

and resolving the velocity derivative along the co-ordinates, we shall represent the solution in the following form, by adopting the method of finite differences:

$$\frac{\partial^2 v}{\partial x^2} = \frac{1}{h^2}[v_{i-1}(t) + v_{i+1}(t) - 2v_i(t)].$$

Then equation (4.70) can be represented in the form of an ordinary differential equation with respect to time for the ith point.

$$\frac{\partial^2 v}{\partial t^2} = \frac{d}{dt}\beta[v_{i-1}(t)+v_{i+1}(t)-2v_i(t)]+\alpha[v_{i-1}(t)+v_{i+1}(t)-2v_i(t)], \quad \ldots (4.71)$$

where an additional term β is introduced, $\beta = 4\eta/(3\rho h^2)$.

The structural network of ABM reproducing a solution to equation (4.71), can be constructed according to the approximation circuit (see Secs. 4.4.1 and 4.4.2) in the form of a more complex chain, each cell of which (Fig. 57, b) should consist of two integrators (1 and 2), a summator (3) generating the second derivative of velocity of rock particle relative to co-ordinates by the method of finite differences, a summator (4) generating the first term in the right side of equation (4.71), along with the block of multiplication (6) and summator (3), summator (5) transforming v into $2\alpha v_1$, which then is fed to the summator (3). The first term of the right side of equation (4.71) is fed to integrator (1) in the first cell of summator (4),

$$\beta\frac{d}{dt}[v_0(t) + v_2(t) - 2v_1(t)],$$

while the second term of this part of the equation goes through summator (3)

$$\alpha[v_0(t) + v_2(t) - 2v_1(t)],$$

forming with integrator (1) the structure of the equation being solved at a distance $\Delta x = h$ relative to the first nodal point, i.e., solving for velocity v_i.

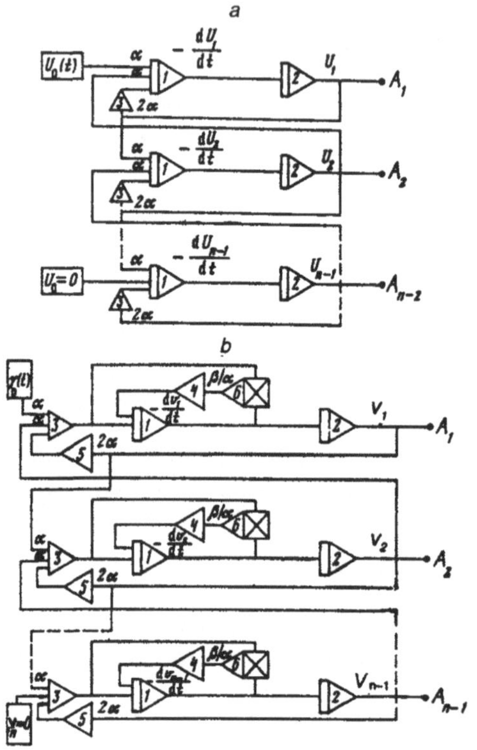

Fig. 57. Structural networks of ABM, generating solutions to a unidimensional wave equation describing (a) elastic and (b) elastoplastic movement of rock induced by blasting. u_i—displacement in ith rock particle; v_i—velocity of ith rock particle.

A similar operation is conducted by the second cell of the chain relative to the second nodal point (velocity v_2) right up to $(n-1)$-th nodal point (velocity v_{n-1}), giving at the exit of cells in the chain of the ABM network, even A_i corresponding to magnitudes of velocities. Here, just as in the first example at a given boundary, from which time is reckoned with respect to propagation of wave through rock, the magnitude of initial velocity $v_0(t)$ is assigned, while at a considerable distance from this boundary $x \to \infty (i \to n)$ the magnitude of velocity v_n is taken as equal to zero.

Literature Cited

1. Adushkin, V.V., Kozyk, B.B. and Rakishev, V.R. 1976. Issledovanie dinamiki dvizheniya massiva pri vzryvanii ustupov skvazhinnymi zaryadami [Investigations of the dynamics of massif movement during bench blasting using blasthole charges]. In: Razrushenie i deformirovanie tverdoi sredy vzryvom. *Vzryvnoe Delo*, 76(33): 98–109.

2. Borovikov, V.A. 1980. Razvitie gazovoi polosti pri vzryve tsilindricheskogo zaryada v gornoi porode [Development of a gaseous cavity during blast of a cylindrical charge in rocks]. *FTPRPI*, 6: 40–52.

3. Borovikov, V.A., Vanyagin, I.F. and Tsirel, S.F. 1983. Metodika raschyeta koeffitsienta sblizheniya pri ustupnoi otboike [Methodology for computing the proximity coefficient in bench blasting]. *Gornyi Zhurnal*, 8: 45–50.

4. Borovikov, V.A. and Vanyagin, I.F. 1985. Tekhnika i tekhnologiya vzryvnykh rabot [Techniques and Technology of Blasting Operations]. Izd.-vo LGI, Leningrad.

5. Vanyagin, I.F. 1977. Osnovy modelirovaniya protsessov razrusheniya gornykh porod [Fundamentals of Simulating the Processes of Rock Breakage by Blasting]. Izd.-vo LGI, Leningrad.

6. Vanyagin, I.F. and Borovikov, V.A. 1980. Modelirovanie deistviya vzryva pri razrushenii gornykh porod [Simulating Blast Effect in Rock Breakage]. Izd.-vo LGI, Leningrad.

7. Vanyagin, I.F. and Borovikov, V.A. 1984. Fizicheskoe modelirovanie deistviya vzryva i protsessa razrusheniya gornykh porod vzryvom [Physical Modelling of Blast Effect and Process of Rock Breakage by Blasting]. Izd.-vo LGI, Leningrad.

8. Borovikov, V.A., Vanyagin, I.F., Laikhansuren, B. and Belyatskii, V.P. 1983. Zakonomernosti zatukhaniya volny napryazhenii pri prokhozhdenii cherez treshchinu [Laws of attenuation of stress wave when passing through a crack]. In: Seismika promyshlennykh vzryvov. *Vzryvnoe Delo*, 85(42): 52–60.

9. Nasonov, I.D. 1978. Modelirovaniye gornykh protsessov [Modelling the Mining Processes]. Nedra, Moscow.

10. Karpunov, E.G., Misnik, Yu.M. and Nefedov, M.A. 1977. Metodika izmerenii parametrov seismovzryvnykh voln v gornykh porodakh p'eizoakcelerometrami [Methodology for measuring parameters of blast-induced seismic waves in rocks using piezoaccelerometers]. FTPRPI, 5: 103–108.

11. Pavlov, M.M. and Christoforov, B.P. 1976. Elektromagnitnaya metodika izmereniya polya massovykh skorostei pri vzryve tsilindricheskikh zaryadov [Electromagnetic method for measuring the field of particle velocities in a blast of cylindrical charges]. *Vzryvnoe Delo*, 76(33): 196–200.

12. Sedov, L.I. 1970. Mekhanika sploshnoi sredy [Mechanics of a Solid Medium]. Nauka, Moscow, vol. 2.

13. Tarasov, V.S. 1976. Analogovye vychislitel'nye mashiny [Analogue Computers]. Izd.-vo LGI, Leningrad.

14. Urmaev, A.S. 1978. Osnovy modelirovaniya na ABM [Principles of Modelling on an Analogue Computer]. Nauka, Moscow.

Printed in India